Ecological Relationships
of Plants and Animals

Ecological Relationships of Plants and Animals

HENRY F. HOWE
University of Illinois at Chicago

LYNN C. WESTLEY
University of Iowa

New York   Oxford
OXFORD UNIVERSITY PRESS

Oxford University Press

Oxford   New York   Toronto
Delhi   Bombay   Calcutta   Madras   Karachi
Petaling Jaya   Singapore   Hong Kong   Tokyo
Nairobi   Dar es Salaam   Cape Town
Melbourne   Auckland

and associated companies in
Berlin   Ibadan

Library of Congress Cataloging-in-Publication Data
Howe, Henry F.
Ecological relationships of plants and animals.
Bibliography: p.   Includes index.
1. Ecology.   1. Westley, Lynn C.   II. Title.
QH541.H65   1988   574.5   87-5800
ISBN 0-19-504431-2
ISBN 0-19-506314-7

2 4 6 8 10 9 7 5 3 1

Printed in the United States of America
on acid-free paper

To our parents

Preface

Ecological relationships of higher plants and animals are universal, of fundamental importance, and paradoxical. Few introductory ecology texts fail to discuss the trophic pyramid or overlook the impact that herbivores have on natural and agricultural communities, but students are rarely equipped to answer fundamental questions. How can an Earth teeming with plant-eating animals, ranging in size from aphids to elephants, be so green? Plant-eating animals kill more plants than drought or logging, yet they do not consume them all. Why not? The answer must be found in evolutionary adjustments of plant-eating animals to a formidable array of structural and chemical defenses evolved to repel them. Consider another question. Do plants really need animals that pollinate their flowers and disperse their seeds? A variety of devices attract flower-visiting insects, birds, and bats, and an equally impressive array of fruits appeal to fruit-eating birds and mammals. But what difference do these mutualisms make? What happens to tropical plant communities when fruit-eating toucans and monkeys go into the stewpot, as often happens in developing countries? The answers are now being found, but they are not being conveyed to college students. Even though 20–30% of the publications in ecology journals now concern herbivory, pollination, or seed dispersal, these topics receive little attention in ecology texts. There is a widening chasm between what is known in ecology, and what is taught.

Ecological Relationships of Plants and Animals is a text for college students and first-year graduates in agricultural entomology, botany, ecology, forestry, population biology, wildlife management, and zoology. The book is intended to be the primary text in an undergraduate course in plant and animal interactions or coevolution, or as one of three or four short texts in an introductory sequence in population biology and ecology. The book is also short enough to supplement a general ecology course. Until now, most of the fascinating phenomena, issues, and debates in plant and animal ecology have been reserved for advanced graduate students. This is unnecessary. The concepts are straightforward and require only the background in biology, chemistry, and mathematics that most students have by the end of their first year in college. Graduates and professionals may find this an easy reference for a wide variety of issues in herbivory and mutualism, but we have written it for undergraduates with only a good course in introductory biology. Our goal is to bridge the gap between the reality of a rapidly expanding discipline in ecology, and the inadequate representation of that field to the next generation of biologists.

We present the ecology of terrestrial plant and animal interactions as a coherent discipline. Even with minimal attention to marine herbivory, itself a diverse field worthy of a book, the task of maintaining coherence has not been easy. Interest in plant and animal relationships has surged over the last 15 years, but the effects of this surge are uneven. Agriculture, forestry, and tradition have spawned immense literatures on herbivory and pollination, while such topics as seed dispersal, the genetics of coevolution, and the relevance of plant and animal relationships to community ecology have received far less attention than they deserve. We balance coverage of familiar topics with promotion of important but lesser known subjects. Our hope is that a balanced use of a wide range of examples will give college students a stimulating array of things to think about and ultimately do, whether in their backyards, a cattle range, or in the far reaches of a tropical rainforest.

Iowa City
April 1987

H. F. H.
L.C.W.

Acknowledgments

We wish to express our gratitude to the many individuals who helped us write this book. May Berenbaum, Blaine Biedermann, Robert Black, Phyllis Coley, Paul Feeny, Kathleen Keeler, Judith Smallwood, Kathleen Troyer, Sara Via, and David Wiemer read parts of the manuscript and offered numerous comments on content and style. Marikay Klein provided the graphical illustration, and Todd Erickson and Robin Roseman the freehand drawings and insets.

We regret that our necessarily eclectic treatment has omitted far more examples than could be included. The book will evolve as the field evolves. We welcome suggestions from readers on content, approach, and style.

CONTENTS

Ecological Relationships
of Plants and Animals

I

EVOLUTIONARY ECOLOGY

The green world around us presents a paradox. Plants use energy from the sun to fix carbon, and all animal life ultimately depends on use of energy stored by plants. Yet animals do not eat *all* plants. Caterpillars remind us that they have the potential to destroy our green world when they occasionally denude entire forests. But they usually do not destroy forests. Somehow most plants, most of the time, avoid the worst consequences of mandibles and teeth.

Charles Darwin provided the key to questions borne of this paradox in *On the Origin of Species*. His central theme was the theory of evolutionary adaptation by natural selection. Darwin argued that plants and animals best fitted to their environments leave more offspring than other members of their species and, consequently, pass their inherited advantages on to future generations. Darwin even provided insight into the nature of plant defenses and animal ability to cope with those defenses. He noted that sheep of different breeds have different susceptibilities to plant poisons. The fittest plant-eating animals have an inherited ability to cope with those plant defenses, and therefore to prosper on foods that their competitors cannot eat.

Darwin used this concept of evolutionary adaptation to explore other plant and animal relationships. The sexual function of flowers was not understood 130 years ago. Darwin correctly inferred that plants advertise nectar rewards to insects that carry their pollen. Nor was it understood at that time why plants so often produce berries useful as food to birds. Darwin argued correctly that seeds dispersed by birds had a better chance of survival than those that were not dispersed. Darwin's theme of adaptive adjustment by natural selection opened the door to an exciting exploration of the ecological relationships of plants and animals.

In this century the contemporary incarnation of Darwinism, evolutionary ecology, has greatly clarified fundamental issues in the study of adaptation, and has helped biologists understand pressing practical issues. Modern tools from biochemistry, genetics, and physiology allow us to apply Darwin's reasoning to phenomena that Darwin might have pondered, but could never have understood. For instance, it is no surprise to an evolutionary ecologist that insects quickly evolve resistance to insecticides. Long evolutionary history has given insects the ability to detoxify a myriad of natural plant poisons, and the potential to evolve resistance to artificial toxins similar to those with which they can naturally cope. Effective pest control, by people or plants, often requires chemicals unlike those to which a particular insect has already evolved resistance.

A general theory is an explanatory synthesis. Successful tests of hypotheses from the theory of adaptive evolution by natural selection strengthen its explanatory power. Unsuccessful tests provide the incentive for further research to explain anomalous discoveries, and thereby bring theory into line with reality. Evolutionary ecology provides a theme for investigating relationships between animals and plants. Its successes elucidate nature; its failures show where there is special potential for creative insight.

1

Adaptation and Natural Selection

The next time you spend a few minutes sitting on a log in a woodland in Spring, look at what is going on between your feet. Are those ants carrying anything? Chances are good that the leaf litter around you is alive with ants that are struggling with what appear to be bits of plant debris. Look more closely. Those flecks of debris could be seeds (Fig. 1-1). Follow the ants to their nest, which is perhaps in the log on which you are sitting. If the wood is decayed, a poke at it may reveal a store of tiny seeds that seem to match those of wildflowers around you. A practiced eye might spot discarded seeds around the entrance of the nest, or even minute bouquets of seedlings that reveal the locations of abandoned ant nests. You may be sitting on top of an ecological interaction that is common from the quiet woodlots of Europe and North America to the vast saltbush scrub of Australia. Did you ever know that ants collected seeds, or wonder why?

Observing nature requires only keen senses. Interpreting it requires an approach to unlocking the riddles that nature offers. Why do those ants carry seeds? Why do the ants carry those seeds and not others? Are the ants and herbs good for each other? An idle observation on a warm spring day could

Fig. 1-1 Ant *(Formica podzolica)* collecting a seed of a violet *(Viola nuttallii).* Note that ants manipulate seeds by the elaiosome, which they later eat. A mature violet is in the background. After Beattie (1985). Copyright © 1985 Cambridge University Press, reproduced with permission.

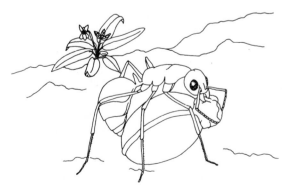

easily lead to a program of discovery on at least three levels: natural history, adaptive function, and evolutionary ecology. Each has something to contribute to understanding nature, whether the subject is seed carrying by ants or any other relationship of animals and plants.

NATURAL HISTORY

Natural history is the description of nature, often from casual or unsystematic observations. It involves finding answers to "which and what" questions. Which ant species gather seeds? What do they do with them? Which herbaceous plant species produce seeds that ants like, and which are ignored? Which plants seem to survive only on ant mounds, or former ant mounds? What happens to herb populations if ants are absent? What happens to ant populations if certain herb species are absent?

"Which and what" questions satisfy curiosity about ant and plant life-styles in one place and time. But they usually have no special organization or theme. They are endless. The list in the previous paragraph could be doubled or tripled in a few minutes. A lifetime might be required to provide the answers. In the end, what would there be except an exhaustive descriptive natural history, without explanatory power?

Descriptive natural history is not itself science, although it often provides the background necessary for scientific investigation. Charles Darwin was a first-class naturalist from childhood, but only became a scientist after worldwide travel and careful reflection gave him a conceptual scheme for observations and experiments. What distinguished Darwin's *On the Origin of Species* (Darwin, 1859) from so many other nineteenth-century treatises on nature was a theme that explained the origin of different kinds of animals and plants without calling upon supernatural forces. The theme, adaptive evolution by natural selection, was so powerful that it also superseded aimless natural history as a rationale and method in the study of nature. With one powerful book, Darwin relegated "which and what" questions to backstage in the theater of science.

ADAPTATION BY NATURAL SELECTION

Adaptation is the match of form and function to the environment. **Natural selection** is the differential reproduction of inherited traits. Differential reproduction of useful qualities allows some parents to contribute a disproportionate share of descendants to future generations. As selection culls individuals ill-suited to their environments and permits those better suited to proliferate, a population becomes molded by its environment.

Naturalists before Darwin recognized that form and function seemed to fit each kind of animal and plant to its circumstances of life. A deer nibbles buds

Table 1-1 Essentials of Darwinian natural selection

Conditions

1. Individual organisms differ in structure, color, behavior, and physiology (**phenotypes vary**).
2. Some variants produce more offspring than others (**fitness varies**).
3. Offspring resemble parents (**fitness is inherited**).

Result

4. The environment favors the reproduction of parents with some inherited traits over those with others (**natural selection occurs**).

with delicate jaws, a turkey crushes hard seeds with a muscular gizzard. Grasses of windswept steppes rely on air currents to carry pollen from their flowers to those of others of their kind. Trees in the stillness of the rainforest use insects to do the same job. Some early naturalists even suspected that similar animals or plants shared a common descent, suggesting evolution from one kind of animal or plant to another. But no one had a clear idea of how this could happen. Darwin provided a synthesis of everyday observations in what we now call the general theory of adaptive evolution by natural selection.

Darwin's primary contribution, the mechanism of natural selection, is disarmingly simple (Table 1-1). First, he noticed that all living things are variable. Any population of plants or animals has members that differ in size, shape, resistance to disease, ability to endure stress, fertility, growth rate, and many other characteristics. Second, Darwin noticed that some parents produce more offspring than others of their species. Any species has the capacity to fill the Earth with its descendants, but none do. In a world of limited resources, most parents fail to leave descendants. Some leave many. Third, Darwin recognized the age-old adage "Like begets like." Animals and plants tend to resemble their parents. He reasoned that parents with an inherited advantage tend to be the ones that produce more offspring than others. In an evolutionary sense, fitness refers to the contribution of a parent to subsequent generations.

Natural selection seems self-evident now, but in the nineteenth century it took a disturbing insight to expose its force. In 1838, Darwin was impressed by the troubling prediction of economist Thomas Malthus that starvation was the inevitable fate of humankind because people produced more children than food to feed them. Whether or not this is always true for humans, Darwin saw that organisms generally have the *capacity* to produce far more descendants than the Earth can support. Every species has the potential to overwhelm the Earth, but none do.

Darwin calculated that a single pair of slow-breeding elephants could produce 15 million descendants in a few hundred years, if each mother produced six youngsters that survived to breed during her 90-year lifetime. Obviously this does not happen. There are probably a million elephants on Earth today, and the two living species have been breeding for hundreds of thousands of years. Most female elephants must produce substantially fewer than six surviving young in their lifetimes.

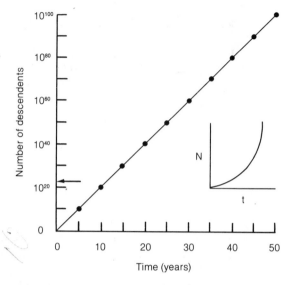

Fig. 1-2 Potential increase in an ant-dispersed violet. The calculation assumes that 50 seeds of one adult, and all 50 seeds of each descendant, survive and reproduce each year. The arrow indicates the number of stars in the known universe. Note the log scale on the vertical axis. Inset: Exponential increase on a numerical scale.

The ant-dispersed violet *(Viola nuttallii)* is an example of a more prolific organism that produces about 50 seeds each year (Fig. 1-2). Imagine that each seed survives to become a breeding adult the next year, so that every adult is replaced by 50 new adults generation after generation. An equation for exponential increase can predict how many violets will exist at any specified future time. The effective birth rate b would be 50. The population size N at time t would be

$$N_t = b^t N_0$$

where N_0 represents the original population of one parent with 50 seeds ($N_0 = 1$). This unchecked growth is exponential (inset in Fig. 1-2). In only 25 generations, the number of violet plants would exceed the number of stars in the known universe (10^{22}). If each plant weighed a gram, within 50 generations all of the descendants of one parent would weigh 60 times more than the Earth! The *potential* for increase is astronomical, but of course this potential is never realized for long. Most young violets, like other young plants or animals, die of disease, crowding, or lack of light or shelter, are eaten by animals, or simply do not reproduce. Only an elite fraction of parents contribute their genes to future generations because only an elite fraction of the total offspring in a population both survive and reproduce. And some of these elite "reproductives" leave far more surviving offspring than others.

Darwin reasoned that virtually any inherited characteristic assisting in this metaphorical "struggle for existence" will be favored, because such traits determine which offspring survive and consequently which parents contribute the most to future generations. The key elements of natural selection (Table 1-1) can operate in any population of animals or plants. But the force of selection may be especially strong in prolific organisms with a high capacity for natural increase. Most of the plants and insects discussed in this book are just such organisms.

How does the theme of natural selection help organize our perceptions of nature? Think of violets and ants again. If natural selection favors seed collection for both the plant and the ant, those parent plants with seeds attractive to ants should, on average, produce more adult offspring than those parents whose seeds are not planted by ants. Ant colonies that collect seeds but do not eat them should, for some reason, fare better than colonies that either consume the seeds or ignore them. Both of these predictions are testable hypotheses that can be used to tease apart the reasons why plants and animals interact as they do.

How can a blind process like natural selection produce apparently perfect adjustments of form and function? Darwin disputed other intellectual leaders of his time by arguing that design by a perfect creator was not necessary to explain apparently perfect adaptations. Natural populations responding to environmental challenges over many generations could produce the apparent perfection in the eagle's eye or the hummingbird's wing. In the example of ants and herbaceous plants, it is not difficult to imagine ants collecting seeds for food, and herbs originally defending seeds with stocks of toxins. It is a short step for plants to provide a tasty seed coat, contrasted with a distasteful or even poisonous seed, to transform destructive seed-collecting into a potentially effective means of dispersal.

At the same time, adaptation is a tricky concept that is not always easy to apply. One problem, discussed below, is that the genetics of adaptations is often obscure. The other problem is that it is easy to think of plausible adaptive explanations that are simply wrong. For instance, an insect biologist might guess that a toxin in herb seeds is a defense against seed-eating ants, while a microbiologist might think of the same chemical as a defense against fungal pathogens. A plant biochemist, unfamiliar with ecology, might view the poisonous chemical as an incidental by-product of plant metabolism. Because several adaptive explanations are often plausible, no single interpretation should be accepted at face value unless alternatives have been excluded.

With such potential for mischief, is adaptation a useful idea? One critic, the geneticist Richard Lewontin (1978), simply concludes that the concept of adaptation is problematic, but inescapable. More than a century after Darwin, the concept of adaptive evolution by natural selection is the best way to explain the penguin's flipper and the elephant's trunk. It is equally powerful in explaining why plants produce poisons, and how animals as different as grasshoppers and buffalo have adapted to prosper on those poisonous plants.

EVOLUTIONARY ECOLOGY

Evolutionary ecology is the study of the ways in which species adapt to their surroundings. Evolutionary ecology uses modern ecological methods and a modern understanding of adaptive evolution to provide both answers and a common theme to "which and what" questions from natural history. Quantitative statistical methods allow biologists to decide which answers can be accepted and which cannot, and modern genetic theory provides a far more complete background for interpreting sources, directions, and rates of evolution than Darwin could have imagined. Evolutionary ecology is really Darwinian inference with modern tools.

Ecology

Ecology is the study of the relationships of animals and plants to their nonliving and living environments. Ecologists make a distinction between two different sorts of environmental challenges: abiotic and biotic.

The **abiotic environment** consists of physical variables, such as temperature and moisture, that determine where species can survive and reproduce. Each physical variable is continuous, and every animal and plant species reproduces best along some part of each physical continuum. Climate determines the broad spectrum of physical conditions to which a species is subjected. But local conditions often vary widely within a region that has the same overall climate. A species of small bee capable of living and reproducing in a region might be unable to fight wind and heat stress on exposed hillsides, but may prosper in valleys or even in the shade and protection of large boulders. Most plants are sedentary, except as seeds. A seedling or adult bush is entirely dependent on the texture and richness of the soil, the amount of light that it can intercept for use in photosynthesis, and the amount of water provided by the climate and local drainage patterns.

The **biotic environment** includes all of the other plants, animals, and microorganisms with which a particular species interacts in nature. For an animal, a minimal list includes disease organisms, food species, symbiotic gut organisms, competitors, and predators. Additionally, a bird or mouse might have special needs, such as tree cavities for cover or vine bark for nests. Plants are influenced by disease organisms, competitors, symbiotic root microbes, herbivores that eat parts of plants, and predators that eat and kill whole plants. Most higher plants also interact with animals that pollinate flowers, disperse seeds, or both. Of course, the biotic environment of an individual organism includes members of its own species.

The concepts of abiotic and biotic environments are linked, because both simultaneously affect every plant and animal on Earth. At a fundamental level, plants use light energy from the sun to convert carbon dioxide gas into carbon-based sugar molecules that can either serve as energy storage for the plants, or

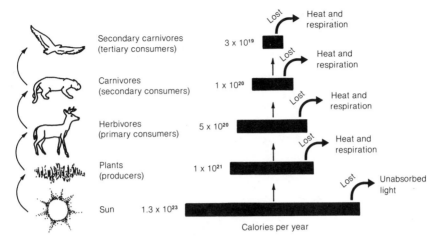

Fig. 1-3 The energy pyramid of a land-based community. Note that a large proportion of the energy is lost between trophic levels. Trophic levels are pictured and labeled on the left. After Sherman and Sherman (1986).

as structural building blocks of cell walls. In turn, plants are necessary for all animal life. Animals feed directly on plants or eat animals that feed on plants (Fig. 1-3). These relationships of energy transfer may be represented by a pyramid of **trophic levels,** with plant producers at the bottom and animal consumers that eat plants or other animals at successively higher levels. The number of organisms and their weight in organic matter (biomass) decreases with each successive level of the trophic pyramid because energy consumed by metabolism is dissipated at each level, leaving less for consumers than producers, and less for secondary consumers than for primary consumers (Fig. 1-3). In this most fundamental scheme, physical variables such as light, temperature, and moisture determine how effectively plants capture the sun's energy, while biotic interactions between animals and plants determine how effectively plants use the energy that they have captured for their own purposes.

Energy transfer is not a sufficient descriptor of biological interactions, however, because critical interactions often involve little energy transfer. Mice eat seeds, and foxes eat mice. Little energy is transferred, but the seed-eating mice may limit tree reproduction. Foxes may even influence tree abundance by influencing mouse abundance. Likewise, many plants require insects for pollination, and provide pollen that bees and flies use as food. Pollen represents a minuscule fraction of the trophic pyramid, yet the success or failure of pollen transfer determines the abundance and rarity of many plant and animal species of a field or forest.

As early as 1917, the pioneering naturalist Joseph Grinnell defined such different roles as distinctive "ecological niches" that could be specified for any species. More recently, Evelyn Hutchinson (1959) defined the **ecological niche** as a "mutidimensional hypervolume in space and time." What he meant was

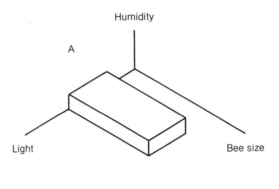

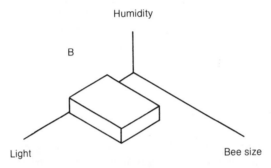

Fig. 1-4 Schematic representation of a plant niche with the three resource dimensions humidity, light, and pollinator (bee) size. (A) Fundamental niche, assuming no competition from other species. Each point within the box represents a set of conditions in which the species can survive and reproduce. (B) Realized niche, assuming large bees are co-opted by another plant species. This plant survives and reproduces in a more limited niche volume than it would in the absence of the competitor for large pollinators.

that every resource needed by an animal or plant can be represented on a continuous axis of a graph. The **fundamental niche** of a species is defined as the set of resources and physical factors that permit survival and reproduction. The fundamental niche of a plant species that uses three resources can be represented by a cube, which illustrates all of the combinations of dimensions in which the plant can survive and reproduce (Fig. 1-4A). If some competitor expropriates some part of a resource continuum, the fundamental niche is restricted to a smaller **realized niche** (Fig. 1-4B). In the hypothetical case illustrated here, a competing plant expropriates large bees, thereby limiting the realized niche of its competitor shown here to the small end of the bee size continuum.

Of course, all species use more than three resources. An actual fundamental or realized niche might include dozens of resource dimensions. Moreover, plants and animals often use each other as resources. Nonetheless, the simplified concept of the multidimensional niche formalizes four important insights into ecology: (1) many resources and physical variables influence the distribution and abundance of a species, (2) resources vary in space and time, (3) resources critical to some species are unimportant to others, and (4) competition does not occur unless the presence of one species limits the availability of resources to another.

The objective of ecology is to tease apart the biotic and abiotic factors that directly or indirectly influence the distribution and abundance of species. Evolutionary ecology further attempts to determine how ecological variables mold the adaptive fit of organisms to their environments.

Genetics

Genetics is the study of heredity. Even an elementary understanding of genetics shows what a tremendous advantage biologists now have over Darwin in discovering how organisms adapt to their world, or why they fail. Unlike Darwin, biologists now know how hereditary information is passed on, the genetic consequences of sexual reproduction, and the relationship between heredity and structure and function.

Particulate inheritance explains one of Darwin's greatest riddles; why many parental characteristics are faithfully passed on to offspring rather than blended together. Genes (cistrons in some texts) are sequences of deoxyribonucleic acid (DNA) bases that code for unique amino acid sequences, or polypeptides. Polypeptides, alone or in combination with others, form protein molecules. Protein molecules are structural components of cells, or enzymes that regulate cell function. They link the genetic constitution of an organism (genotype) and its appearance (phenotype) to selective agents in nature.

Genes are arranged linearly along chromosomes. In higher organisms, one chromosome is donated by the "mother" and one by the "father" (in plants, the pollen and egg are often from the same individual). For each active gene position or locus (Latin for location) on a chromosome, there is some identifiable polypeptide product. A gene locus may be invariant, or be represented by alternative forms called alleles. A locus with two or more alleles is polymorphic. Different alleles often produce slightly or dramatically different polypeptides, which may help or hinder cell function. Although an *individual* organism has only two chromosomes of a homologous pair, and consequently a maximum of two alleles for a given locus, a large population of individuals may contain up to a dozen or more alleles for a given locus.

The importance of different alleles is that the slightly or dramatically different proteins for which they code may influence the ability of their bearers to survive or reproduce. For instance, allele differences at two loci determine whether clover *(Trifolium repens)* is capable of producing cyanide gas (see Chapter 5). Cyanide-producing (cyanogenic) clover plants grow more slowly than members of the same species that do not possess the cyanide genotypes. Slow growth is more than compensated for in meadows where clover-eating insects and snails are common, however, because most insects and snails avoid cyanogenic plants.

Tremendous hereditary variation exists in most natural populations because of mutation and recombination. Mutations are random changes in chromosome or gene structure. Chromosomal mutuations involve duplication, loss, fusion, or rearrangement of chromosomes or parts of chromosomes. Gene loci may move from one chromosome or part of a chromosome to another location, without being lost. Point mutations change the base sequence in DNA molecules, and consequently may produce different alleles from originally identical genes. When organisms reproduce sexually through meiosis and mating, recombination mixes alleles at respective loci on different chromosomes.

Mutation and recombination ensure genetic diversity. Mutation, the ultimate source of genetic variation, occurs at rates of 10^{-5} or 10^{-6} per locus per gamete. Each human has 10^5 gene loci, and consequently carries an average of two alleles not present in either parent. Over innumerable generations, mutation has produced alternative alleles at 15 to 70% of the gene loci of most animals and plants. High genetic polymorphism creates tremendous evolutionary potential. In a diploid organism with two copies of each chromosome, the potential number of genetically unique gametes (sperm or eggs) produced by sexual recombination is 2^n, where n is the number of loci with two different alleles at each. If there are 2,000 such loci in an average individual (likely for humans), the number of different kinds of gametes is 2^{2000} or about 10^{600}. In a population of hundreds, thousands, or millions of possible parents, the potential genetic variability of their offspring is virtually infinite.

Without knowing the mechanism of hereditary transfer, Darwin could not understand the link between the hereditary particle (gene) and the structural or functional characteristic upon which selection acts. Biologists now know that natural selection acts directly upon phenotypes, and only indirectly upon genotypes (see Appendix I). Traits heavily influenced by alternative alleles at one locus are defined as **single locus traits.** In such traits one allele (a dominant allele) may completely mask the phenotype of its opposite allele (a recessive allele). This is complete dominance. Other single locus traits show only partial masking of recessive phenotypes (incomplete dominance), or no masking effect at all (no dominance). Traits influenced by alleles at many loci, each with a small effect on phenotype, are **polygenic traits.** Polygenic traits may be influenced by some loci showing dominance, by some that interact with each other, and by some that have an incremental additive effect on phenotype.

Natural selection acts somewhat differently on single locus and polygenic traits. Alternative alleles often produce distinctive phenotypes in single locus traits, and consequently natural selection can easily discriminate between phenotypes with different effects on fitness. Moreover, identical phenotypes often indicate identical genotypes, except when complete dominance masks a recessive allele. Variation in polygenic traits is continuous rather than discrete, so selection cannot discriminate as easily between gene differences as it can in the single locus case. Moreover, many genotypes may produce the same phenotype. For instance, millions of people on Earth share your height, which is a polygenic phenotype. Unless you have an identical twin, none are your exact genotype.

Finally, geneticists have begun to understand why selection on one trait may produce unexpected changes in other traits. For instance, virtually all mutations affecting fur color in guinea pigs also affect viability, litter size, or behavior (Wright, 1980). Genes with **pleiotropic effects** influence more than one phenotype. This may occur because protein products have a direct influence on two or more phenotypes, or because their polypeptide products interact with those of other gene loci (epistasis). Gene loci may also be connected physically on the same chromosome (genetic linkage). When selection strongly favors or discriminates against the phenotype of an allele *A* at one locus, it may coincidently produce a change in frequency of the phenotypes influenced by an allele B at *another* locus linked to *A*. Pleiotropy and linkage add uncertainty to the evolutionary process because the total phenotypic result of selection can be unpredictable.

Modern genetics presents evolutionary ecologists with issues that Darwin never faced. Some important traits in plant and animal interactions, such as plant production of simple chemical defenses, are due to single locus effects. Many other traits, such as the size of flower and fruit crops or the intensity of animal foraging for plants, may be influenced by many genes. An interpretation of the apparently adaptive responses of animals and plants to each other requires biologists to keep in mind several lessons from heredity: (1) a trait may not respond to selection because gene differences do not influence its phenotypic expression; (2) selection may be strong or weak (Appendix I); (3) genetic dominance or interactions among loci may slow or speed response to selection; (4) a trait that might seem significant to a biologist may be an incidental by-product of selection on a very different characteristic of the organism; (5) a trait of interest to a biologist might reflect a past history of selection rather than ongoing adaptation. Genetic constraints on adaptation are as important as the genetic mechanisms by which adaptation occurs.

SUMMARY

The concept of evolutionary adaptation has itself evolved since Darwin's day. Biologists now know that many different selective pressures simultaneously act on every species of animal and plant. Biologists also know that the response of a species to selection can be simple and predictable if traits are strongly influenced by single genes, or complex and less certain if traits are polygenic and influenced by genes with pleiotropic effects. A fossil record full of species that no longer exist (Chapter 9) is an impressive testimony to the complex and sometimes cumbersome nature of the genetic machinery that makes evolutionary adaptation possible. Genetic systems that cannot adapt to the environment become extinct. The ecology of plant and animal relationships is best understood through a thorough knowledge of biotic interactions in nature and, where possible, through some understanding of the genetic background of traits that mediate these interactions.

STUDY QUESTIONS

1. Often people think of the "struggle for existence" as literal combat between competitors or predators and prey. What are some real or hypothetical examples of natural selection in which there is no contact between competing individuals?
2. Imagine the differences in reproductive output of progeny lines of elephants averaging six and two reproductive offspring per parent, respectively. Calculate those differences for 5 generations. Repeat the exercise for 30 generations.
3. Assume that a giant sequoia tree *(Sequoiadendron giganteum)* produces 2 million seeds each year for 500 years. If two seedlings colonized a previously unoccupied continent, how many descendants would there be in 100 generations if 0.3 of 1% of the lifetime reproductive output of each tree produced breeding young? How many sequoia seedlings do you think actually survive?
4. George Williams (1975) argues that selection is more intense on organisms with many offspring than on those with few. Does this mean that sequoia trees are evolving much faster than Dutchman's breeches? Than humans? Why or why not?

SUGGESTED READING

Douglas Futuyma (1986) presents a synthetic overview of evolution in *Evolutionary Biology*, while Charles Krebs (1985) does the same for ecology in his *Ecology*. Brandon and Burian (1984) provide a distinguished collection of essays on the intricacies of natural selection and adaptation in *Genes, Organisms, Populations: Controversies over the Units of Selection*. No one could do better than to start with The Master, Charles Darwin, in *On the Origin of Species* (facsimile of the 1859 edition).

2

Testing Hypotheses
in Evolutionary Ecology

Nonscientists often have the impression that scientific knowledge grows by accretion; like a calcium deposit accumulating precipitate, science progresses, or
perhaps just enlarges, as more and more facts are discovered. Practicing scientists know that actual scientific progress is both more idiosyncratic and more
systematic than this accretion model implies. Idiosyncracy enters when a scientist chooses one of several guiding themes as an approach to his or her research. Then a systematic program of discovery, tailored by the standards of
the research area, favors some approaches to the exclusion of others.

Effective scientists narrow their attention to questions that help them resolve
important issues. Important issues are defined by a **scientific theory,** which is
a coherent and logically interrelated set of ideas. Scientists then pose **alternative hypotheses,** which are logical statements, interpretations, or predictions
derived from theory. A hypothesis states what "should" happen if a particular
interpretation of theory is correct. To determine correctness, scientists measure
whatever a particular hypothesis requires, and analyze the measurements. If
findings agree with predictions, the hypothesis survives until the next test. If
facts do not come out as predicted, the hypothesis is rejected. If many hypotheses are rejected, the theory that produced them must be modified to accommodate the facts. Sometimes negative evidence is so overwhelming that
the theory itself must be abandoned in favor of one with more explanatory
power. The most productive science tests hypotheses that discriminate between
two or more plausible interpretations.

What are the important issues in the evolutionary ecology of plant and animal interactions? At the most general level, the Darwinian notion of adaptation
by natural selection is the guiding theme in evolutionary ecology. An evolutionary ecologist is always on the alert for some bit of natural history that
appears to contradict adaptation by natural selection. Darwin himself posed the
ultimate alternative hypothesis that would overturn his theory:

> If it could be proved that any part of the structure of any one species had been
> formed for the exclusive good of another species, it would annihilate my theory,
> for such could not have been produced through natural selection (Darwin, 1859,
> p. 201).

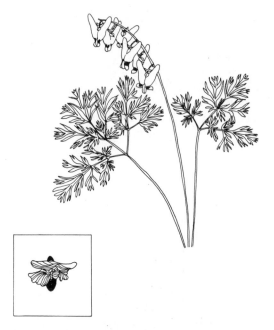

Fig. 2-1 Flower of Dutchman's breeches *(Dicentra cucullaria)*. Inset: seed showing the edible elaiosome, here a folded outgrowth of the seed coat much sought by ants as food. Drawing by R. Roseman.

This is a strong challenge. The Darwinian notion of adaptation by natural selection is a powerful scientific theory because it is both plausible and open to a fundamental test. Darwin gives the conditions necessary to "annihilate" the theory with one critical discovery. If it could be shown, for instance, that wild plants and animals evolved to feed and clothe people, rather than to propagate themselves, Darwin's theory would collapse.

At another level, the theory of evolution by natural selection holds within it an almost infinite array of lesser theories, each with its own alternative hypotheses. These may be used to explore the intricacies of the general theory of adaptation by natural selection without testing the general theory itself. In this chapter we will explore a series of tests of alternative interpretations of a real and, at the time of writing, uninvestigated phenomenon in the ecology of plant and animal relationships.

Suppose the idle moments in a springtime forest envisioned in the first chapter led you to notice ants carrying seeds of a common herbaceous plant (herb) of the woodland floor (Fig. 2-1). A field guide quickly identifies the plant as the common Dutchman's breeches *(Dicentra cucullaria)*, so named because of its odd pantaloon-like flowers. By good fortune an entomologist friend, or your college library (e.g., Creighton, 1950), helps you identify a common ant, *Aphaenogaster rudis*. What are the ants doing with those seeds?

As working hypothesis, suppose that the ant is a seed predator that kills the seed without helping the plant in any way. **Seed predation** means that an animal, in this case an insect, feeds on and kills a seed just as a fox kills a rabbit. The ants eat seeds to obtain energy that helps the ant colonies survive and reproduce. The herb gains nothing, but loses potential offspring to marauding ants. This hypothesis presumes no prior knowledge of either these particular ants and plants or of others like them. It simply assumes that ants, like many other insects, slugs, snails, reptiles, birds, and mammals, eat plants.

An alternative hypothesis might be that ants and plants share a **mutualism,** in which each species benefits the other. But mutualism would be a far-fetched hypothesis without prior information. Mutualisms are less common than antagonistic relationships between species. How does an evolutionary ecologist distinguish these two alternatives?

COMPARATIVE METHOD

The **comparative method** tests evolutionary hypotheses by matching evidence from related animals or plants against predictions from general theory. For instance, an evolutionary ecologist might suppose that plant-eating insects become adapted to similar poisons in related plants, and in fact are limited to those plant families. The first step in testing this hypothesis is to survey host preferences of insects for different plant groups. Paul Ehrlich and Peter Raven (1964) used published host records from the entomological literature to test this hypothesis. Consistent with the prediction of host specialization between insect families, they found that some families of butterflies rear their young on certain plant families, others rear their young on others. Could a similar approach illuminate ant and seed interactions?

The first step of a comparative approach is to test the working hypothesis, that *Aphaenogaster rudis* eats *Dicentra cucullaria* seeds, by reading about related phenomena in a college or university library. A good study of these two species might exist. But most situations in nature have never been studied explicitly. It is far more likely that a library search will show (1) whether ants (perhaps even this species) eat seeds, and (2) whether herb seeds (perhaps even *Dicentra* seeds) are consumed by ants. The best way to start will be with recent secondary references, such as books or reviews, and abstracting references (*Biological Abstracts, Botanical Record, Current Contents, Zoological Record*). These will point to primary references in original journal articles.

We start with ants. Articles in prominent journals document the extraordinary impact that seed-eating ants of the genera *Pheidole, Pogonomyrmex,* and *Vermessor* have on desert vegetation (e.g., Brown et al., 1979). Such ants consume so many seeds that they compete with desert rodents without providing any obvious service to the plant. However these papers do not mention *Aphaenogaster,* and *Dicentra* grows in woodlands, not deserts. Further sleuthing reveals the *Aphaenogaster* species often do disperse violet *(Viola)* seeds in eastern deciduous forests of the United States (Culver and Beattie, 1978). It and several other ant genera (e.g., *Lasius, Myrmica*) seem to take many seeds to nests, and discard them apparently unharmed. But other *Aphaenogaster* species eat seeds. Comparative evidence suggests that some mutual benefit (mutualism) may accrue to both ant and plant, but the possibility of seed predation is still alive.

Next, we read about the plants. No detailed study of ant dispersal of *Dicentra* seeds exists, but some articles contain intriguing comments. Some authors simply state that our species is probably ant-dispersed (Schemske et al., 1978). Why would they think that? The bibliography of that article leads to a review by Berg (1969), who describes *Dicentra* as a good example of a genus endowed with **elaiosomes,** or oil bodies on the seed coats, that are apparently attractive to many ants (Fig. 2-1, inset). A few more minutes in the library reveal that the woodlot phenomenon is widespread. Handel and his co-workers (1981) find that nearly a third of the herb species in one forest in the eastern United States have elaiosomes. These herb species make up over half of the individual plants in the woods. The list includes such familiar plants as wild ginger *(Asarum),* sedge *(Carex),* spring beauty *(Claytonia),* bloodroot *(Sanguinaria),* violets *(Viola),* and even Dutchman's breeches *(Dicentra).* Our plant seems to be one among many species thought to be ant-disseminated.

Based on known behavior of *Aphaenogaster rudis* ants and the presence of an elaiosome on *Dicentra cucullaria* seeds, we now have reason to try to distinguish the hypotheses of mutualism and seed predation. Comparative evidence has given us a strong clue that the original hypothesis of seed predation is wrong. But nowhere have we seen direct evidence that this ant species eats only the elaiosome of seeds of this particular *Dicentra* species. The ants might still eat the seeds and kill them. Common sense adds a third possibility; seeds in ant nests may not be eaten, but may die there because conditions are poor for germination and growth. Perhaps the ants discard otherwise healthy seeds in the wrong kind of place. We need direct evidence.

The comparative method yields only partial answers because differences among species confound interpretations. Even related animals or plants differ in many ways that have nothing to do with a particular ecological or evolutionary hypothesis. But the comparative method is often the surest means of posing appropriate questions to ask of direct observations or field experiments.

OBSERVATIONS

An afternoon in the woodlot led to an hour or two in the library, and information from the library leads us back to the woodlot. The three alternative hypotheses now are (1) ants are seed predators, (2) ants remove seeds for elaiosomes but are not effective dispersal agents, and (3) ants disperse seeds. What do we look for?

First, we watch the ants. They appear to pick up seeds of several species, not just those of *Dicentra cucullaria*. When they return to their nests, they carry the seeds inside. We know from the literature that some seed-dispersing ants discard seeds around the entrance of the nests, while others keep them inside. These ants keep *Dicentra* seeds inside.

Second, we dig up nests, most of which are under flat stones or are among the leaves on the forest floor. There are seed caches in these shallow nests, and the seeds appear intact, except that the white elaiosomes have been chewed off. Seed dispersal is certainly possible, but perhaps the seeds simply die in the nest. Seedlings cannot get light under a rock, and we do not see seedlings in active nests. The hypothesis of seed predation still has not been excluded; perhaps we have only found the uneaten seeds in nests. *Aphaenogaster* might eat elaiosomes first, seeds later.

Third, we use another clue from the library to tighten the case. Judith Small-wood and David Culver (1979) discovered that this ant species regularly moves its nest. In fact, a nest site is occupied for an average of only 23 days before it is abandoned (Smallwood, 1982). The question now is whether ants leave the seeds to germinate, eat them, or take them to a new nest. A tedious search in the leaf litter might turn up a little cluster of seedlings that could show the location of an abandoned nest. But we cannot be sure of either the species of seedlings or the identity of an old nest site. The best approach is to mark each of several active nests so that we can keep track of them, and let the ants go about their business for a few weeks.

Direct field observation has given us a better understanding of ant and herb natural history, and seems to point toward mutualism. But the issue is still not resolved. One could sift an enormous amount of leaf litter without *distinguishing* alternative hypotheses. Even if marked nests are eventually abandoned, and seedlings grow up from them, seedlings might still do better elsewhere in the leaf litter than in old ant nests. One must alter the system, and see if the experiments can clear up ambiguities.

EXPERIMENTS

Experiments control some facets of nature so that uncontrolled aspects can be explored without confounding interpretations. Controlled conditions artificially reduce the confusion in nature so that interesting phenomena can be studied. Experiments can be difficult to devise. Without direct field observations or a solid grasp of the literature, someone might hold the important variables constant and explore trivial ones. But a successful experiment can resolve in a few days what might take years to accomplish by direct observation.

The key information that would resolve the issues of seed predation, seed waste, or mutualism is a determination of whether or not the per-capita survival of a seed taken by *Aphaenogaster rudis* is higher than the per-capita survival of a seed not taken by this ant. An attempt to mark individual seeds will probably fail. Seeds are tiny, and ants react to most paints. Another approach is to take *Dicentra* seeds to a nearby patch of woods where the plant happens to be absent. We risk trying to plant seeds where the seedlings cannot grow. But use of such a site would allow some confidence in the source of *Dicentra* seedlings. Obviously we should not draw conclusions from germination of only one or two seeds. The inexorable compounding of numbers that might occur in any population tells us that virtually all *Dicentra* seeds must die. What we should look for is evidence that more die if they are not taken by *Aphaenogaster* than if they are. In order to make that comparison, we must use a substantial number of seeds.

A reasonable planting scheme must be devised. We might place a control sample of seeds (perhaps 100) on the surface and the same number under the surface at a reasonable ''ant nest depth'' (known from looking at nests) under and away from *Dicentra* plants. We could do the same on top of and inside of *Aphaenogaster* nests. To minimize error, the procedure should be replicated. Fifteen times is easy enough. One hundred replications would probably ensure that the whole experiment need not be repeated in another season.

What will be the outcome of this experiment? We, the authors, do not know. Perhaps all seeds on the surface will be eaten by mice. Perhaps those that we scatter on top of ant nests will survive as well as those planted in real or simulated ant nests. Perhaps only those in ant nests survive, or only those away from ant nests. Perhaps any buried seed is equally likely to survive. Perhaps all buried seeds survive, but seedlings in abandoned ant nests are larger and healthier because the soil has more nutrients. We have chosen an interaction which neither we nor anyone else seems to have explored. Might you?

This exercise is a partial exploration of this ant and herb interaction. At the time of writing, only Culver and Beattie (1978, 1980) have probed the issue in this fashion, using *Aphaenogaster* and violets *(Viola)* in West Virginia, and other ant and violet species in southern England. In North America, seedlings were 10 times as likely to emerge from ant nests as elsewhere, and much the same held for some English violets. If *Aphaenogaster rudis* treats *Dicentra cucullaria* as it does some violets, seeds will be far more likely to germinate in old ant nests than elsewhere. In such an event, mutualism would certainly be the likeliest explanation for our casual observation on a warm spring day.

INTERPRETING RESULTS

A balanced use of comparative, observational, and experimental evidence maximizes the power of the scientific method. With all three tools, any problem in plant and animal interactions can be approached with confidence. The comparative method shows how particular species resemble or differ from others. Observations sometimes resolve questions and always help us devise sensible experiments. Sensible experiments actually distinguish alternatives. We then interpret the results. One kind of interpretation is immediate; it tells us whether our observations are consistent enough to be believed. A more wide-ranging interpretation puts the study into some general context which furthers our overall understanding of nature.

Analysis

Modern ecological analysis is based on explicit statistical procedures. If all patterns in nature were clear-cut, such procedures would not be necessary. But in ecology most patterns vary, either because they are inherently variable or because many factors influence them. This is not the place to insert a course in statistics. But an example will show how ecologists decide that some hypotheses are likely to be true, while others are not.

Hypothesis testing is a means of separating alternative arguments. Some decisions are easy. Suppose an ant nest contained only fragments of seeds. You would know that the ant was a seed predator. Suppose you found that elaiosomes had been chewed off, leaving intact seeds, but the seeds and seedlings were moldy. You would decide that the ants were not seed predators, but were destructive to the plant because they left seeds in unfavorable locations. With such obvious results, statistical analysis would not be necessary.

But suppose, as is usually the case, that results are not clear-cut. Based on a solid understanding of the natural history of violets, Culver and Beattie (1980) decided that it was important to discover whether seeds had equal germination under four conditions. They planted 100 violet seeds in each of four treatments (Table 2-1). Seeds with elaiosomes were planted 2 cm deep singly (treatment 1) and in clumps (treatment 2). The scientists removed elaiosomes from 100 more seeds, and planted them in clumps (treatment 3). Then they allowed ants to pick up unaltered seeds and carry them to nests (treatment 4). Most seeds failed to germinate in all treatments (Table 2-1), although more ''ant-planted'' seeds survived than others. Is the success of ''ant-planted'' seeds really disproportionate? Here a chi-square test tells us whether the emergence patterns shown might have occurred by chance (Zar, 1984). The test produces a chi-square value of 27.0, with a **significance value** of $p < .001$. The significance value

Table 2-1 Seedling emergence of English violets planted in different treatments[a]

Planting Treatment	Seedlings Emerging	Seedlings Not Emerging
1. Unaltered and single	12	88
2. Unaltered in clumps	7	93
3. Elaiosome removed in clumps	9	91
4. Planted by ants	30	70

Source: Data from Culver and Beattie (1980).

[a]The χ-square statistic $\chi^2 = \Sigma(\text{observed} - \text{expected})^2/\text{expected} = 27.0$ with 3 degrees of freedom ([4 rows-1][2 columns-1]). Observed values are shown. Expected values are calculated as the products of (column total × row total)/total of all observations. For row 1 under seedlings emerging, the value is $(58 \times 100)/400 = 14.5$. See Siegel (1956).

suggests that the germination pattern shown might occur by chance only 1 out of 1,000 times. The pattern shown is almost certainly due to something other than chance. Ants help the seeds survive through the germination process.

This is only one kind of statistical test. Others are mentioned in subsequent chapters. Sometimes the hypothesis tested requires a comparison of distributions, which was the case above. Sometimes tests determine whether mean values differ (e.g., average number of bee visits to one plant compared with average bee visits to another), or whether some samples are more variable than others. A statistical correlation determines the likelihood that two variables (e.g., number of bee visits and number of flowers per plant) are or are not really associated with each other.

Big Pictures

Even an apparently specialized scientific result can help many people understand nature. An experiment should interest people other than the handful of specialists on *Aphaenogaster* or *Dicentra*. Results should be integrated within a context broader than a narrow sliver of insight might at first suggest.

For instance, an ant and herb interaction can be both a confirmation of general Darwinian principles, and an exploration of the rather poorly understood mutualisms between ants and plants. Because we are working *within* the Darwinian conceptual framework (a paradigm of Kuhn, 1970), the study is likely to be a general confirmation of an already well-confirmed generality, adaptation by an inferred process of natural selection. Either seed predation or mutualism could be interpreted in this light. The well-known phenomenon of ant dispersal of other seeds with eliaosomes precludes the interpretation that herbaceous plants produce nutritious food bodies and sacrifice seeds for the good of ants rather than as a means of procreation (Beattie, 1985). This is not the kind of test that would, if it failed to show mutualism, prevail over Darwin's challenge that was reprinted at the beginning of the chapter.

On another level, this test might provide an exacting probe of the nature of ant and plant mutualism. A demonstration of *either* seed predation *or* mutualism would allow a useful exploration of the details of poorly understood interactions between ants and herbaceous plants. A demonstration of mutualism between *Dicentra cucullaria* and *Aphaenogaster rudis* would concur with other studies reported in the literature. This positive result would provide strong additional evidence that *Aphaenogaster* has evolved an affinity for various herb seeds, and that *Dicentra* has evolved a mechanism for using ants as dispersal agents.

If *Dicentra* seeds fail to survive in *Aphaenogaster* nests, our results are an exception to the *Aphaenogaster rudis* pattern. This would also be an interesting result. If a certain "seed-disperser" ant and an "ant-dispersed" herb fail to show a mutualistic relationship, what determines which species cooperate and which do not? If some species of ants are better dispersal agents than others, how might herbs encourage some ant species and discourage others? If elaiosomes of some herb species are better food for ants than others, what determines which ants take the best seeds, and which take the leftovers? Each of these questions leads to new questions. Even in this corner of a local woodlot, the more one learns, the more one needs to find out.

The adaptive context is only one general framework. We might be interested in a community hypothesis. For instance, we might ask whether plants compete for ants, or vice versa. Or we might wonder whether ants influence the distribution and abundance of plants, or vice versa. An attempt to fit a narrow result into a wider context is both instructive to other people, and it leads to formulation of new hypotheses. Much as science is not a collection of facts in textbooks, scientific knowledge does not simply grow as an accumulation of isolated results.

SUMMARY

Science is a process of exploration, not a collection of observations. The scientific method uses comparative, observational, and experimental approaches to match data with predictions from alternative hypotheses. Statistical analysis helps scientists decide which hypotheses are accepted and which are rejected. Further interpretation in the broader contexts of community or evolutionary ecology helps put apparently isolated discoveries in perspective, and often leads to new hypotheses.

STUDY QUESTIONS

1. The example developed in this chapter assumes that both the plant *(Dicentra cucullaria)* and the ant *(Aphaenogaster rudis)* observed in a local woodlot could be identified easily. In reality, local wildflowers are easy to identify, but insects must sometimes be sent away to a specialist in a university entomology department for identification. How would you proceed with your study if you could not investigate the species of ant in your college or university library until some weeks after the season was over?
2. An ecologist asks for a large grant to *prove* that mites damage spruce forests. Why would you be suspicious of this scientist's objectivity?
3. A botanist notices that a cardinal eats berries, and crushes all of the seeds. Repeated observations and experiments verify the impression that cardinals are "seed predators." The botanist then argues that nutritious fruit flesh is an anachronism, no longer advantageous in nature, because modern birds digest both the pulp and the seed. What step in the scientific method has the botanist neglected?
4. Table 2-1 gives the results of four "treatments" in a seedling emergence

experiment. What other treatments might have been tried? What confounding factors might make the treatments in Table 2-1 difficult to interpret?

SUGGESTED READING

An especially provocative analysis of how science works is Thomas Kuhn's (1970) classic, *The Structure of Scientific Revolutions*. A wider variety of views are given in essays in the anthology *Criticism and the Growth of Knowledge,* edited by Imre Lakatos and Alan Musgrave (1970). Ashley Montagu (1984) edits a fine anthology about conflicts between science and pseudoscience in *Science and Creationism*. Clear, readable, and self-contained elementary texts in statistics include Sidney Siegel's (1956) *Nonparametric Statistics* and Jerrold Zar's (1984) *Biostatistical Analysis*. For students with a new-found interest in ants, Andrew Beattie's (1985) *The Evolutionary Ecology of Ant-Plant Mutualisms* is a must.

II

A GREEN EARTH AND
ITS ENEMIES: HERBIVORY

Life as we know it exists because plants use energy from the sun to convert carbon dioxide and water into organic molecules that animals eat. A paradox is that apparently edible greenery covers most of the land area of the Earth, despite the fact that millions of organisms eat plants. Why is the Earth so green? Are bacteria, beetles, and buffalo so limited by disease or predators that they cannot make full use of an infinite food supply? Or do the plants ensure their own defense?

The next three chapters probe the mechanics, dynamics, and coevolution of interactions between plants and animals that eat plants. Plants use a variety of devices to protect their roots, stems, leaves, and seeds, ranging from cellulose roughage that slows digestion to exotic amino acids that interfere with protein formation in animal cells. Plant-eating animals call on an ancient heritage of counter-defenses, ranging from behavioral avoidance to digestive chemicals that dismantle lethal plant molecules. Plants and their enemies are locked into a silent struggle of adaptation and counteradaptation that neither can afford to lose.

Biologists should not take these struggles lightly. Human civilizations are entirely dependent on plants for food and fiber. Whatever threatens corn, tomatoes, cotton, or the grasses that sheep eat, also threatens us.

3
Plant Defense
and Animal Offense

Land plants first emerged from lakes, rivers, and swamps 450 million years ago, and it is probably no accident that the animals came ashore right after them (Chapter 9). **Herbivory,** the consumption of plants by animals, evolved in ancient lakes and oceans, and undoubtedly was part of primeval land life. A long history has given plants time to evolve an enormous array of mechanical and chemical defenses against animals that eat them (herbivores). The apparently infinite supply of green foods in fields, forests, and prairies is not as palatable as it looks. Much of it is genuinely poisonous. Herbivores, on the other hand, require plant material for sustenance and are also capable of evolution. Four hundred and fifty million years of experience have given herbivores a complex arsenal of adaptations for breaking plant spines, digesting plant fibers, and detoxifying plant poisons.

Natural selection is the engine that powers an inexorable and intricate contest between plants and herbivores. In every woodlot, slough, meadow, and rainforest on the planet, plants use mechanical and chemical tools to survive and reproduce in the company of herbivores that could, if unrestrained, annihilate them. Herbivores likewise struggle to neutralize plant defenses, and at the same time outreproduce competitors of their own or other species. Relationships between plants and herbivores are not static. A walk in a desert, field, or forest reveals plants that look healthy alongside others that have lost the battle against their enemies (Fig. 3-1). Darwin's metaphorical "struggle for existence" has turned against a cactus demolished by wild pigs in the Arizona desert, a seedling skeletonized by leaf-mining insects, or a seed riddled by weevil grubs. Such failures of defenses illustrate the potential for annihilation of plants by herbivores. Occasional plagues of defoliating insects realize this destructive potential for entire ecosystems.

PLANT ARSENALS

Plant defenses include (1) mechanical protection on the surface of the plant, (2) complex polymers or silica crystals that reduce plant digestiblity to animals, and (3) plant toxins that kill or repel herbivores at very low concentrations.

Fig. 3-1 Herbivore damage. (A) A spiny prickly pear *(Opuntia)* eaten by small wild pigs called peccaries *(Tayassu tajacu)* in the Sonoran Desert of Arizona. (B) A seedling skeletonized beyond recognition by insects in Panama. (C) A nutmeg *(Virola)* seed riddled with the burrow of a weevil *(Conotrachelus)* grub. Photographs by H. F. Howe.

Mechanical Protection

External structural defenses on stems and leaves discourage animals ranging in size from leafhoppers to bison (Fig. 3–2). Sharp cactus **spines,** which are actually highly modified leaves, repel mammals (note Cooper and Owen-Smith, 1986). Smaller herbivores, such as insects, can be discouraged, injured, or even killed by defenses of stem and leaf epidermis scaled to their diminutive sizes. To an animal that is one or a few millimeters long, a leaf or stem covered with needlelike **trichomes** is a dense tangle of dangerous hooks or spikes. For some plants, **glandular hairs** combine physical and chemical deterrence by impeding and then entrapping insects in sticky and chemically active exudates.

Mechanical protection from small insects or mites is extremely important in many economically important crop plants (Norris and Kogan, 1980). For instance, selecting strains of pubescent (possessing glandular hairs or trichomes) cotton *(Gossypium)* or field beans *(Phaseolus vulgaris)* can make the difference between resistance and susceptibility to locally common pests. Breeding crop resistance to small insects or mites often requires selecting genetic strains that are well-endowed with glandular hairs or sharp trichomes.

Digestibility Reducers

The most important plant defenses are complex polymers or inorganic crystals that make plant cell walls indigestible to animals (Table 3-1). Because these polymers and silica crystals generally inhibit digestion, they are termed **digestibility reducers**. Their effects are dosage dependent or quantitative because the higher their proportion in the diet, the less nutrition animals get from ingesting plant tissues.

Cellulose and **hemicellulose** are complex polysaccharides comprising 80–90% of the dry weight of most plant parts (Fig. 3-3). Omnivores and carnivores digest them incompletely, if at all. Herbivores require numerous digestive modifications that allow slow fermentation by symbiotic microbes. **Pectin** is a closely related polysaccharide that differs in straight-chain structure in $-COOH$ groups where cellulose has $-CH_2OH$ and hemicellulose has a hydrogen. Like hemicellulose, pectin has a branched and twisted three-dimensional structure that assists in its role as a cement between cell walls. Other roles of pectin are controversial. Ruminants digest them easily (Van Soest, 1982), but some insects may not.

Lignins are complex phenolic polymers that stiffen and toughen plant tissues by binding to cellulose or hemicellulose. They are responsible for woodiness in plant tissues and toughen nonwoody tissues in mature leaves and stems. Lignins interfere with digestion by binding to both carbohydrate substrates and digestive enzymes in the animal gut. They are themselves indigestible. Closely associated with lignins are nonphenolic waxes, or **cutins,** that provide the dual functions of water retention on the leaf surface and digestibility reduction.

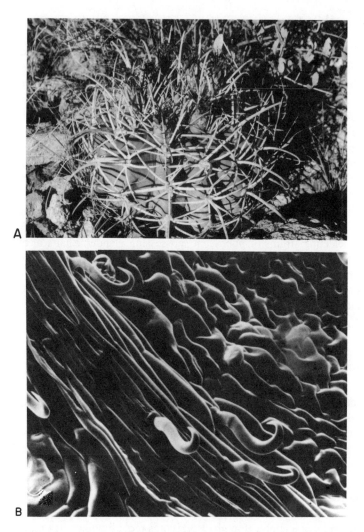

Fig. 3-2 External protection of plants. (A) Cactus spines discourage mammalian browsers. (B) Hooked trichomes on field beans injure small insects. (C) Glandular hairs on a member of the potato family snare mites and aphids, which eventually starve. (A) Photograph by H. F. Howe; (B) photograph by Ward A. Tingey, Cornell University Agricultural Station © American Association for the Advancement of Science; (C) photograph by R. W. Gibson, Rothamsted Experiment Station.

Tannins, like lignins, are polyphenols of seemingly infinite complexity (Fig. 3-3). They differ from lignins in being compartmentalized or free in the plant cytoplasm rather than bound to polysaccharides. Some tannins (e.g., smaller soluble molecules) probably have metabolic roles in plants. Especially important for plant defenses are condensed tannins, which occur in the greater majority of woody plants and in many herbs. By binding to proteins, condensed tannins impede animal digestion of plant tissues by (1) blocking the action of digestive enzymes, (2) binding to proteins being digested, or (3) interfering

C

Fig. 3-2 (Cont'd)

Table 3-1 Selected plant products that reduce herbivory[a,b]

Chemical Group (number identified)	Description	Defensive Role
Quantitative Digestibility Reducers		
Cellulose (1 basic type)	Sugar polymer	Requires gut flora for digestion
Hemicellulose (1 basic type)	Sugar polymer	Requires gut flora for digestion
Lignins (indefinite)	Phenolic polymers	Bind with proteins and carbohydrates
Tannins (indefinite)	Phenolic polymers	Bind with proteins
Silica (1 basic type)	Inorganic crystals	Indigestible
Qualitative Toxins		
Alkaloids (20,000)	Heterocyclic N-containing	Many; some stop DNA and RNA production
Toxic amino acids (260)	Analogues to protein amino acids	Compete with protein amino acids
Cyanogens (23+)	Glycosides that release HCN	Stops mitochondrial respiration
Glucosinolates (80)	N-containing K salts	Many; endocrine disorders
Proteinase inhibitors (indefinite)	Proteins or polypeptides in subunits	Bind with active site of enzymes
Terpenoids (100,000+)	Polymers of C_5 units	Many; some stop respiration

[a] See Rosenthal and Janzen (1979) for a discussion of other secondary compounds.

[b] C, carbon; H, hydrogen; K, potassium; N, nitrogen; HCN, cyanide.

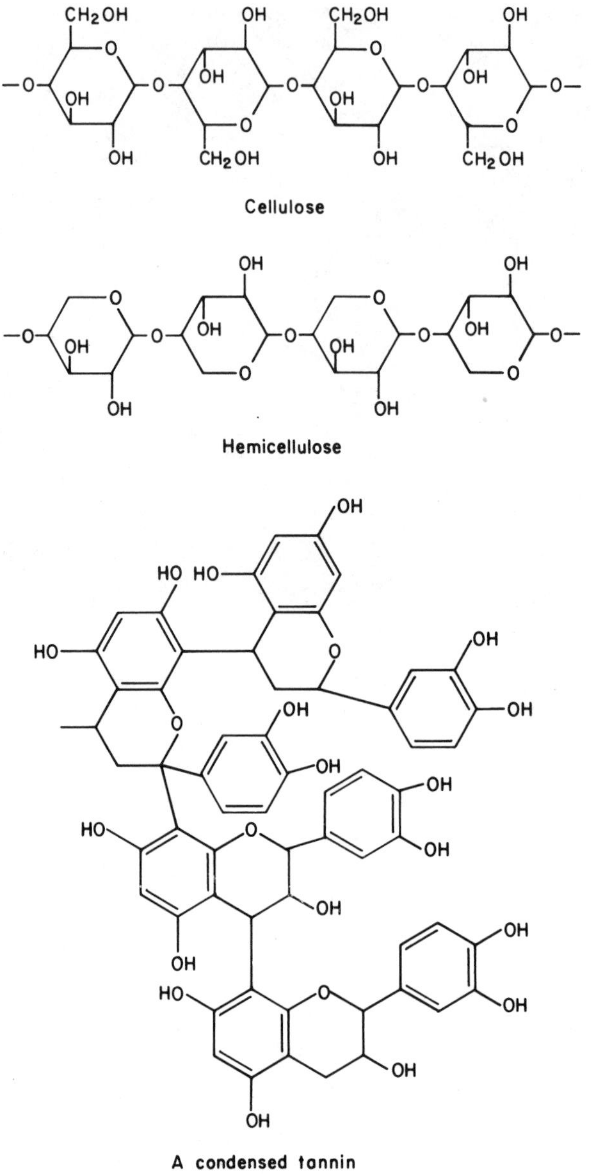

Cellulose

Hemicellulose

A condensed tannin

Fig. 3-3 Digestibility reducers in plants. Cellulose is an unbranched chain; hemicellulose resembles cellulose less than is shown because it is branched and twisted. Condensed tannins may be enormous molecules. After Van Soest (1982).

with protein activity in the gut wall (Van Soest, 1982). Ruminants and other animals that eat large quantities of tannin-rich vegetation often get little nutritional benefit (Robbins et al., 1987). Symptoms of excessive tannin consumption include reduction in growth rate, weight loss, and other symptoms of malnutrition. Extraction of tannins from cell walls improves digestibility for livestock. Agricultural plants bred for animal fodder invariably have lower tannin contents than their wild relatives.

Finally, **silica** is completely indigestible to animals. Silica is an oxide of silicon, one of the most common elements in the Earth's crust. Many plants require silica as a nutrient, using it to strengthen cell walls and, in an unknown way, to assist in carbohydrate metabolism. Grasses and horsetails (*Equisetum*) accumulate silica in the crystalline pattern of opal in quantities that dramatically reduce the digestibility of their mature tissues.

Fossils show that cellulose, hemicellulose, lignins, and silica have provided both support and protection to plants for hundreds of millions of years. It is hardly a coincidence that the most successful structural components of cell walls are difficult or impossible for herbivores to digest.

Toxins

Plant chemicals may be toxic to some animals whether or not they have metabolic functions within plant cells. Toxins with a clearly defensive rather than metabolic role are called secondary compounds, although some authors refer to them as secondary metabolites or allelochemicals. These chemicals have many metabolic sources (Fig. 3-4), and are often selectively produced or stored in vulnerable tissues that have not yet become fortified with lignins or silica, such as developing buds, young leaves, or unripe fruits.

Unlike complex polymers with dosage-dependent effects, simpler secondary

Fig. 3-4 Biosynthetic origins of primary and secondary plant products.

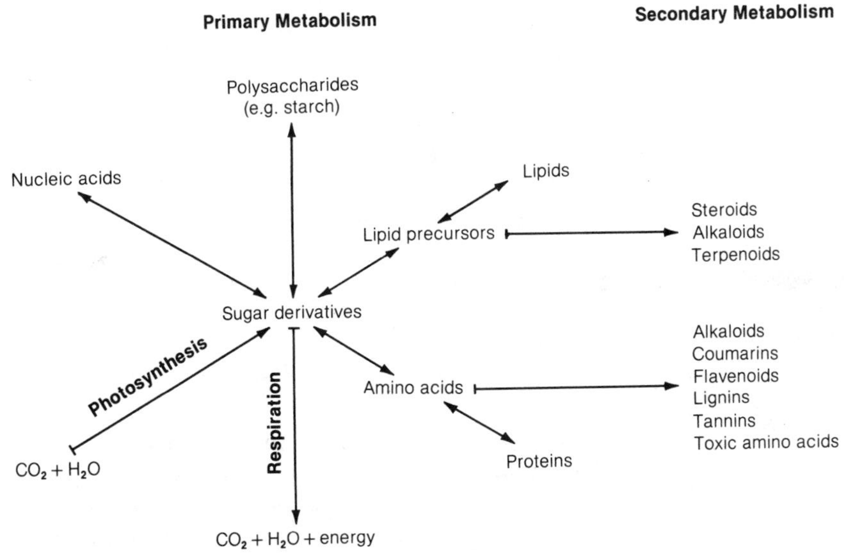

Primary Metabolism   Secondary Metabolism

Polysaccharides (e.g. starch)

Nucleic acids

Lipids

Steroids
Alkaloids
Terpenoids

Lipid precursors

Sugar derivatives

Photosynthesis

Respiration

Amino acids

Alkaloids
Coumarins
Flavenoids
Lignins
Tannins
Toxic amino acids

$CO_2 + H_2O$

Proteins

$CO_2 + H_2O + energy$

compounds often block specific biochemical reactions. Beyond a low thresh-hold, these **qualitative toxins** are poisonous, unless an herbivore has a defense against their special activity. Tens of thousands of qualitative plant toxins exist. A few examples here and in subsequent chapters illustrate their range of structural complexity, chemical activity, and biological utility (Table 3-1).

Alkaloids are heterocyclic molecules containing nitrogen. Approximately 20,000 different alkaloids have been isolated from plant materials, including such well-known compounds as nicotine in tobacco, caffeine in coffee and tea, strychnine in some tropical trees, and the drugs cocaine and morphine (Fig. 3-5). Plant chemists estimate that 15–33% of the dicotyledonous plant families contain alkaloids, with the largest number represented by herbaceous annuals (Robinson, 1979). Alkaloids are especially common in families of the orders containing magnolias (Magnoliales), buttercups (Ranunculales), and gentians (Gentianales), as well as in such familiar flower families as poppies (Papaveraceae), beans (Fabaceae), and asters and sunflowers (Asteraceae). They are virtually absent from cucumbers (Cucurbitales) and from most monocotyledons.

The toxic effects of alkaloids include inhibition of DNA and RNA synthesis (caffeine), inhibition of mitosis (colchicine), stimulation of ribosome breakdown (mescaline), and breakdown of membrane compartmentation (tomatine). Several alkaloids (e.g., physostigmine), like military nerve gases, block the enzyme acetylcholinesterase at nerve synapses.

Terpenoids are among the most biologically important plant products (Mabry and Gill, 1979). All terpenoids have a repeating structure of C_5 isoprene units (Fig. 3–5). Some terpenoids are found in virtually all higher plants because they are integral to plant metabolism. For instance, the plant growth hormone gibberellic acid is a virtually universal terpenoid. Other terpenes have more restricted functional roles. For example, volatile terpenes are aromatic and are often used by plants to attract pollinators.

Terpenes with primarily defensive roles are often restricted to particular plant taxa, indicating evolutionary responses to herbivores that are peculiar to specific plant lineages. The monoterpenes (C^{10}) called the pyrethroids are found in species of *Chrysanthemum* (Asteraceae). Pyrethroids are effective commercial insecticides because they have an almost instantaneous "knockdown effect" on flying insects. They are not especially toxic to mammals. Other monoterpenes are responsible for the resistance of cedar wood *(Thuja plicata)* to attack by termites and beetle larvae, conifer resistance to bark beetles, and the insecticidal properties of catnip (nepetalactone in *Nepeta cataria*). The toxicity of some monoterpenes is due to an inhibition of respiration in mitochondria. For others, the specific mechanism of toxicity is unknown.

Sesquiterpene lactones (C^{15}) are characteristic of sunflowers (Asteraceae), although they also occur in some other families (e.g., Magnoliaceae, Lauraceae, Umbelliferae). Sesquiterpene lactones are found within leaves and stems, or sequestered in surface trichomes and glandular hairs. Many of the 900 known sesquiterpene lactones, often called bitter principles, have impressive toxic effects. For instance, glaucolide-A is found in certain species of sunflower genus

Alkaloids

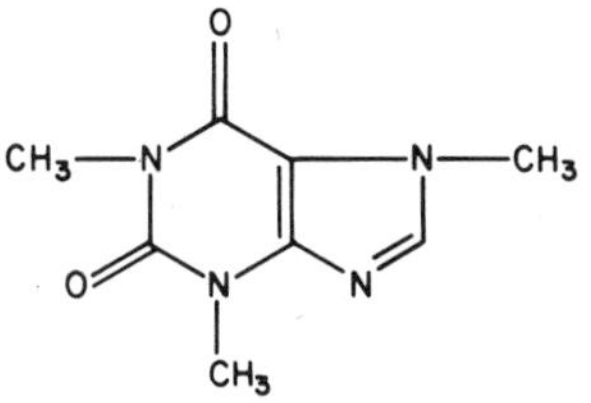

Caffeine

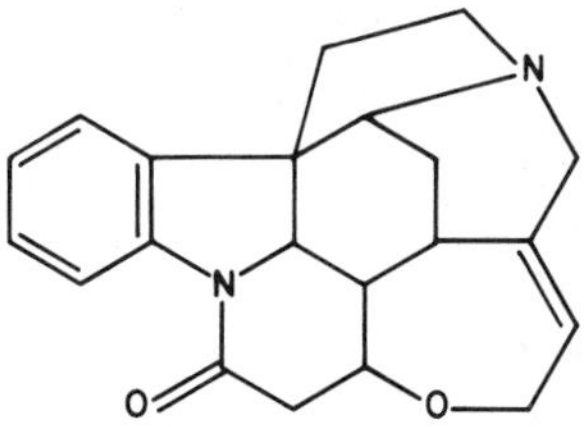

Strychnine

Terpenes

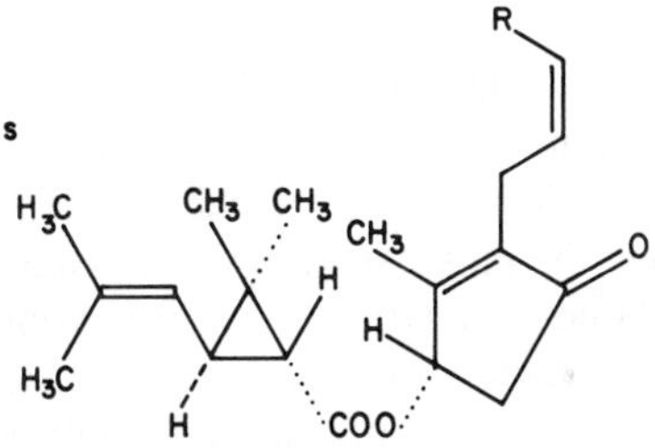

Pyrethroid

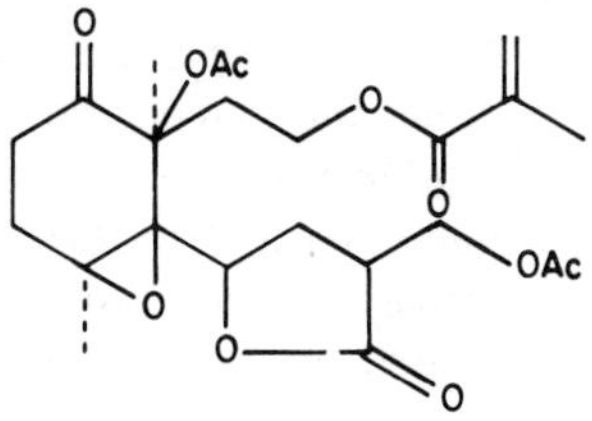

Glaucolide – A

Fig. 3-5 Selected secondary compounds. Top: The alkaloids caffeine from coffee beans *(Coffea)* and strychnine from strychnine fruits *(Strychnos)*. Bottom: The terpenoids pyrethrin from chrysanthemum *(Chrysanthemum)* and glaucolide-A from a sunflower *(Vernonia)*.

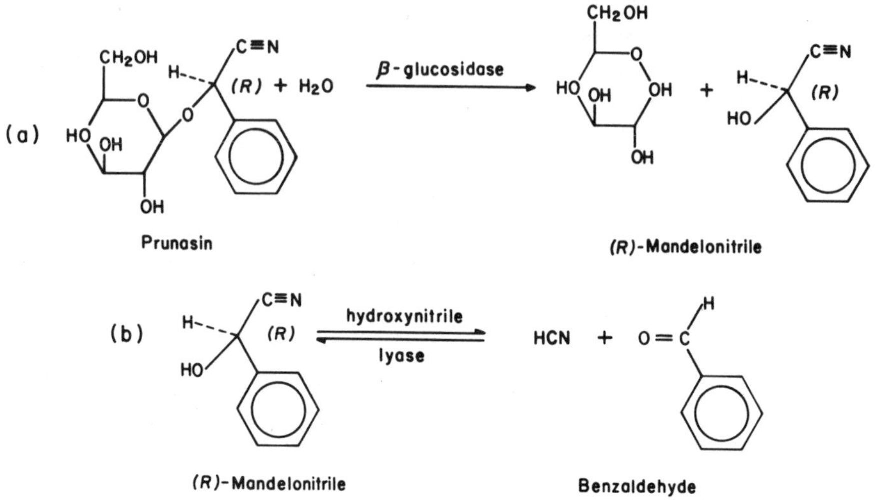

(a)

Prunasin

(R)-Mandelonitrile

(b)

(R)-Mandelonitrile

Benzaldehyde

Fig. 3-6 Cyanide production by damaged cherry leaves *(Prunus)*. After Conn (1979).

Vernonia, but not in others (Burnett el al., 1978). Saddle-backed caterpillars *(Sibine stimulea)* and cottontail rabbits *(Sylvilagus floridanus)* readily eat *Vernonia flaccidifolia,* which lacks glaucolide-A, but reject *Vernonia gigantea,* which has it. Among the most famous plant sesquiterpenoids are ecdysones, or insect molting hormones (Slama, 1979). These mimic the insect hormones that prompt molting after each growth state. Ecdysones are widespread, but are especially important in the defense of such conifers as balsam fir *(Abies balsamea)* and yew *(Taxus baccata).* Insects feeding on plants that produce an analogue of the normal ecdysone produced by their own endocrine systems either fail to molt or fail to develop mature gonads.

Bitter principles are sometimes economically important to ranchers. For instance, as low a dose as 85–100 mg/kg of helenalin or hymenovin can sicken or kill sheep and goats that eat bitter sneezeweed *(Helenium microcephalum)* and bitterweed *(Hymenoxys odorata)* in the United States. Related compounds in *Myoporum deserti,* a common Australian shrub, cause liver and kidney damage in livestock. More complex terpene polymers, the diterpenes (C_{20}) and triterpenes (C_{30}), are also responsible for a wide variety of noxious effects on herbivorous mammals.

Hydrogen cyanide (HCN) is made of hydrogen, carbon, and nitrogen, three of the most common elements in nature. Yet HCN is so toxic to cellular respiration that it cannot exist free in healthy animal or plant cells. As many as 1,000 species of plants sequester cyanide bound to other molecules. It is harmless as long as plant tissues are intact, but can be released when leaves, stems, or seeds are crushed by an herbivore (Conn, 1979). The chemical molecules carrying this simple but lethal toxin are called **cyanogenic glycosides,** which usually consist of a phenol ring, a carbohydrate molecule, and the all-important $C \equiv N$ component (Fig. 3-6). Common families known to contain cyanogenic compounds include roses (Rosaceae), beans (Fabaceae), grasses (Gramineae), aroids (Araceae), sunflowers (Asteraceae), and euphorbias (Euphorbiaceae). Many plants have put this simple toxin to good use with spectacular effects.

How can HCN kill a cow in a dose of 2 mg/kg body weight, or a human in a dose of 3.5 mg/kg body weight? Free HCN gas has a dramatic effect on animals because it blocks the action of cytochrome oxidase, an enzyme critical to cellular respiration. Acute HCN poisoning quickly affects the vertebrate or insect brain, heart, and nervous system, rapidly bringing about stupor, convulsions, and coma. Livestock gorging on sorghum, acacias, or members of the rose family sometimes kill themselves with cyanide poisoning, while even the presence of cyanogenic glycosides prevents insects and other invertebrates from eating some strains of clover. Cooking removes HCN from plant tissues (e.g. almonds) but leaves a toxic residue in highly cyanogenic foods (e.g. cassava).

Constitutive and Inducible Defenses

Defenses may be built into higher plants, or induced by damage from herbivores or disease organisms. **Constitutive defenses** are the permanent protection of a plant species. They include most external mechanical protections like spines or trichomes, as well as the huge variety of chemical compounds that reduce the digestibility of plant tissues or poison herbivores outright (Table 3-1). **Inducible defenses** are responses by individual plants to tissue damage. Proteinase inhibitors and a few other chemical defenses are now known to be inducible, but the list is likely to grow to include many toxins that are presently thought to be constitutive.

Proteinase inhibitors are polypeptides and proteins that block the catalytic activity of proteolytic enzymes in the animal gut by binding to the active site of the enzyme molecule (Ryan, 1979, 1983). Their action is specific to a particular digestive enzyme. Unheated proteinase inhibitors are toxic, but denaturation by cooking converts them to digestible protein. These digestive inhibitors are thought to be extremely widespread in the plant kingdom.

Ryan and his colleagues have worked out the details of induction of proteinase inhibitors in potatoes *(Solanum tuberosum)* and tomatoes *(Lycopersicon esculentum)*. The leaves of each species contain two inhibitors with molecular weights of 41,000 and 23,000, respectively. The larger inhibitor consists of four subunits, the smaller of two. Because the subunits of the two inhibitors differ in amino acid sequence, they appear to have evolved independently.

Leaves of potato and tomato plants respond to chewing by the Colorado potato beetle *(Leptinotarsa decimlineata)* by accumulating the large inhibitor in large quantities and the smaller molecule in smaller quantities in both damaged and undamaged leaves. Leaf damage releases small polysaccharide fragments, called wound hormones, that induce the production and transport of proteinase inhibitors throughout the plant. Within hours, >1% of the soluble protein of the leaf consists of proteinase inhibitors.

Overview

The mechanical, structural, and allelochemical defenses discussed in this book only touch on an immense variety of mechanisms that plants use to preclude or respond to herbivory. Two points need emphasis.

1. Natural-products chemists have named far more secondary compounds than biologists have studied. Information about the ecological effects of most secondary compounds is scarce or nonexistent. The toxicity of an allelochemical is rarely if ever known for more than a tiny fraction of the slugs, insects, birds, mammals, or pathogens that might taste it in nature.
2. A given plant species has several to many protective devices, not just one or two. For instance, a legume might have trichomes as a seedling, tough leaves and lignified wood as an adult, young foliage loaded with cyanogenic glycosides and terpenoids, and seeds provisioned with toxic amino acids and alkaloids. The diversity of allelochemicals in any given species is rarely known, and the overall protective consequences of a total mechanical, structural, and allelochemical defense system have not yet been worked out for any wild plant.

In short, an enormous variety of mechanical and chemical defenses occur in plants. Unhappily, the roles of these complex arrays of defenses in protecting plants from animals are far from thoroughly understood. This is a field ripe for creative synthesis.

EATING PLANTS

Animal enemies of plants range in size from minute aphids to massive elephants (Fig. 3-7). To a plant-eating animal, a world that appears a monotonous green to us is actually a mosaic of palatable, noxious, and deadly patches of foliage or seeds. Daniel Janzen (1978) once commented that "the plant world is not colored green; it is colored morphine, caffeine, tannin, phenol, terpene, canavanine, latex, phytohemagglutinin, oxalic acid, saponin, and L-dopa." To a slug, beetle, or hare, a forest or meadow is a many-textured landscape of scents and tastes, the distinction of which may be a life and death matter. Moreover, one grub's meal is another's poison. Rabbits and saddle-backed caterpillars reject plants containing the terpenoid glaucolide-A (Fig. 3-5), but cabbage loopers *(Trichoplusia ni) prefer* plants with this allelochemical (Mabry and Gill, 1979). Glaucolide-A is not only benign for cabbage loopers, this terpenoid reduces competition from other herbivores that are less able to detoxify it. Few questions in ecology are of more fundamental and practical interest than "Why do different herbivores eat different plants?"

Plant-eating animals have mechanical, biochemical, and behavioral countermeasures to plant defenses (Table 3-2). A few general principles apply to herbivory in animals as different as beetles and bison.

A

B

C

Fig. 3-7 Hervibores differ in size by more than 10 orders of magnitude. (A) Elephants *(Loxodonta)* often exceed 5,000,000 g. Most herbivores are, like the Colorado potato beetles shown in (C), 0.1 to 1.0 g in weight. Small size constrains the diversity and volume of the endosymbiotic community that herbivores maintain, and consequently limits their use of food plants to those that their symbionts and enzyme systems can master. (B) Aphids *(Aphis)* are small (but not the smallest) herbivores at 0.01 g. This minute individual is struggling through glandular hairs on a leaf surface. (A) photograph by L. K. Johnson of the Smithsonian Institution; (B) Photograph by R. W. Gibson of Rothamsted Experiment Station; (C) photograph by Chip Clark of the Smithsonian Institution.

Table 3-2 Common methods of processing plant tissues

Method	Modification (examples)
Fragmentation of intact tissues	Teeth (mammals); beak and/or gizzard (seed-eating birds); chewing mandibles (chewing insects and their larvae)[a]
Breakdown of cellulose and hemicellulose	Expanded gut volume for microbial fermentation (most herbivorous vertebrates and many insects) or intracellular symbionts (many insects)
Detoxification of allelochemicals	Expanded gut for microflora (above); mixed-function oxidase (MFO) systems (vertebrate liver, insect fat body and midgut); perhaps intracellular symbionts (insects)
Avoidance of noxious food plants	Well-developed taste, smell, and/or vision combined with selective foraging behavior (all herbivores)

[a] A piercing beak allows sap-sucking insects (aphids, planthoppers) to avoid consuming noxious plant roughage.

Mechanical Breakdown

The first principle is that animals fragment, or break up, plant tissues so that cell walls are ruptured. This allows digestive secretions access to the protoplasm, and allows gut microflora access to the cell walls.

Tooth structure reflects mammalian feeding habits (Fig. 3-8). Omnivores (e.g., *Homo sapiens*) and herbivores that browse on soft foliage or fruits (e.g., the deer *Odocoelius virginianus*) have **low-crowned teeth** that are useful for grinding meat, foliage, or seeds. Flat cusps of molars are covered with a single layer of hard enamel, which wears away with use. Grazers such as the horse *(Equus caballus)* and African elephant *(Loxodonta africana)* eat abrasive silica-rich grasses using **high-crowned teeth** that wear slowly because several ridges of enamel rather than one thin layer grind tough plant tissues. High-crowned teeth of rabbits grow throughout the life of the animal. Carnivores (e.g., the mink

Fig. 3-8 Dentition of mammalian herbivores. Low-crowned teeth of an omnivorous (e.g., human or pig) or browsing mammal (early horses) wear rapidly once the cap of hard enamel is worn down to the dentine by abrasive vegetation. High-crowned teeth of a grass-eating herbivore (horse or cow) wear slowly because the grinding surfaces expose vertical ridges of enamel. After Romer (1959).

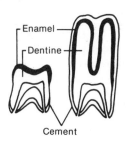

Enamel
Dentine
Cement

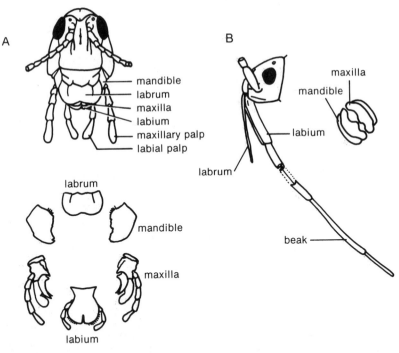

Fig. 3-9 Insect feeding appendages. (A) A grasshopper mouth consists of highly modified appendages for sensing (maxilla), guiding (labrum, labium), and fragmenting (mandible) plant food. (B) The long beak of a seed-sucking milkweed bug is a highly modified appendage including the maxilla and mandible. (A) After Storer (1951); (B) after Borror et al. (1981).

Mustela vison) have sharp-edged carnassial teeth that are useful for shearing meat, but of little use in grinding foliage.

Birds lack teeth, but grind plant material with other organs. For instance, finches like the North American cardinal *(Cardinalis cardinalis)* crush seeds with a thick bill. Turkeys *(Meleagris gallopavo)* break off seed husks with their bills, and use muscular gravel-filled gizzards to crush the kernels. Turkey gizzards generate forces of 3.7 kg/cm^2—enough to crack a hickory nut *(Carya ovata)*. The gizzard is to a turkey what a molar is to a pig.

Long evolutionary experience has given insects a variety of tools for fragmenting plant parts (Fig. 3-9A). Chewing mouthparts have different plans but similar purposes. In some cases, such as the grasshopper shown, both juvenile and adult insects have similar mandibles that carve off and mash pieces of foliage. In other insects the mouthparts and feeding habits of juveniles and adults are entirely different. A swallowtail butterfly caterpillar *(Papilio glaucus)* possesses chewing mouthparts, while an adult swallowtail sucks nectar from flowers.

Some insects avoid the need to fragment plant tissues. True bugs (Hemiptera) and leafhoppers and their relatives (Homoptera) pierce the plant epidermis with a beak and suck plant fluids, thereby avoiding toxic seed coats, trichomes, and toxic epidermal cells. The milkweed bug *Oncopeltus fasciatus* kills milk-weed seeds *(Asclepias)* by piercing them rather than chewing them (Fig. 3-9B). Larval insects that feed on dry seeds often digest the endosperm outside of their bodies with enzyme secretions, and then ingest the "soup."

Microbial Farms

Herbivore digestive tracts harbor a microflora of bacteria, flagellates, and pro-tozoans that synthesize necessary vitamins, break down plant material, and de-toxify allelochemicals through anaerobic (oxygenless) fermentation (Van Soest, 1982). These are essential **microbial symbionts** for plant-eating animals. The microbes strip oxygen from complex organic molecules, thereby reducing them to simple fatty acids, alcohols, and methane that can be excreted or incorpo-rated into useful molecules. In an as yet poorly understood process, different species of microorganisms also detoxify many plant secondary compounds, making otherwise poisonous foods palatable.

Vertebrate guts show extensive structural and functional modification for her-bivory (Van Soest, 1982). Fermentation in mammals may take place before or after the food products enter the stomach (Fig. 3-10). **Ruminants** (e.g., cattle, *Bos*) swallow foliage with minimal chewing, passing vegetable material on to the four-chambered stomach. Cellulose and hemicellulose are degraded into

Fig. 3-10 Mammalian digestive tracts. The sheep is a foregut fermentor with a well-devel-oped rumen and a very long small intestine. The pony is a hindgut fermentor with a heavily sacculated colon. Both are entirely herbivorous. The mink is a carnivore with a simple diges-tive tract and minimal fermentation. After Argenzio (1984). Copyright © 1984 Cornell Uni-versity Press; redrawn with permission.

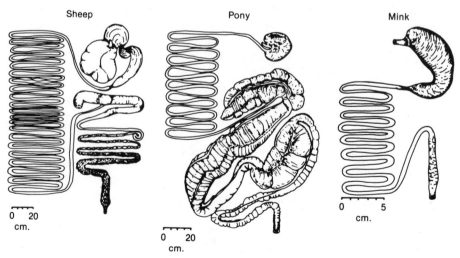

Fig. 3-11 A rumen protozoan, genus *Diplodinium*. Notice the symbiotic bacteria living on the surface of the protozoan; this is an example of two symbionts living within another symbiont, the ruminant (see Vogels et al., 1980). Photograph by C. Davis; from Van Soest (1982).

sugars and starches in the microbe-filled **rumen** (technically, the two-part reticulo-rumen of the stomach). Cattle and sheep chew their cud by regurgitating food, rechewing it, and swallowing it again so new surfaces of cellulose and hemicellulose exposed by repeated abrasion are accessible to rumen microbes (Fig. 3-11). The degraded vegetation in the rumen is passed on to the **omasum** of the stomach, which absorbs some materials and filters undegraded fiber. Finally, fermented material is passed on to the **abomasum** of the stomach and then on to the intestines. The scale of microbial activity is immense in the gut of a plant-eating mammal. A milliliter of sheep rumen fluid contains $16,100 \times 10^6$ bacteria, 10^6 flagellates, and 3.3×10^5 ciliated protozoans. A sheep rumen contains 6 liters of fluid, that of a large cow closer to 80 liters (Parra, 1978).

Nonruminant mammalian herbivores also use fermentation to digest plant fiber. Foregut fermentors such as leaf-eating *Colobus* monkeys have saclike outpocketings of the stomach analogous to the rumen, while hindgut fermentors like the horses have either a large **cecum** (blind sack) or a large **sacculated colon** that houses the microflora (Table 3-3). These alternative fermentation systems are nearly as efficient as the rumen.

Despite the colossal scale of microbial activity in the gut, much plant material cannot be digested (Table 3-4). Success depends largely on (1) the volume of the microbial fluid, (2) the time partially digested food (digesta) is retained

Table 3-3 Gastrointestinal anatomy of selected mammals[a]

Category	Animal
Foregut Fermentors	
Ruminants	Antelope, deer, cattle, sheep
Nonruminants	Colobine monkeys, hamster, hippopotamus, kangaroo, vole
Hindgut Fermentors	
Cecal digesters	Capybara, elephant, rabbit, rat
Sacculated colonic digesters	Horse, New World monkey, man, pig
Unsacculated colonic digesters	Cat, dog

Source: Van Soest (1982).

in the gut, and (3) the proportion of lignified or silicified fiber, or otherwise indigestible material, in the plants eaten. These three factors are related. Large volume of microbial fluid and long retention time allow more complete break-down of lignified or silicified foliage than small volume or short retention times. The ability of an animal to eat well-protected foliage at all may depend on its fluid volume and retention time.

Large mammalian herbivores hold plant digesta longer than smaller species, using their more extensive microbial farms to extract more from each mouthful eaten. For instance, white-tailed deer (*Odocoelius virginianus;* body weight 48–100 kg) retain digesta 45 hours while American bison (*Bison bison;* 450–

Table 3-4 Digestive availability of forage components for ruminants[a]

Component	True Digestibility	Limiting Factor
Completely Available		
Soluble carbohydrate	100%	Intake
Starch	90+	Passage with feces
Organic acids	100	Intake and/or toxicity
Protein	90+	Fermentation
Partly Unavailable		
Cellulose	43–73	Lignification, cutinization, silicification
Hemicellulose	36–79	Lignification, cutinization, silicification
Unavailable		
Lignin	Indigestible	Limits use of cell wall
Cutin	Indigestible	Limits use of cell wall
Silica	Indigestible	Limits use of cell wall

Source: Van Soest (1982).

[a]True digestibility is the balance between food consumed and fecal residues, minus metabolic products.

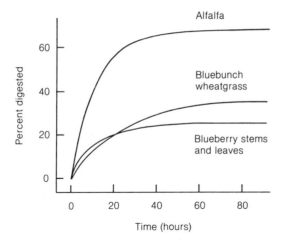

Fig. 3-12 *In vitro* digestion of three mule deer forages, using rumen inocula. Most alfalfa is digested within 40 hours, but heavily silicified wheatgrass and lignified blueberry foliage and twigs are poorly digested after 40 hours. All three forage plants are reduced to indigestible fiber within 60 hours. After Milchunas (1977).

1,350 kg) and domestic cattle (*Bos taurus;* 450–900 kg) take up to 80 hours to process plant fiber (see Robbins, 1983, p. 317). Deer digest up to 56% of the alfalfa fiber that they eat, and cattle digest 70% (Van Soest, 1982). Large nonruminants are almost as efficient. Horses (*Equus caballus;* 350 kg) and elephants (*Loxodonta africana;* 2,800–5,000 kg) digest slightly over 50% of relatively lignified alfalfa fiber. By comparison, an omnivorous human (60 kg) with a much smaller fermentation chamber and a much less complex microbial farm in the intestines digests only 9% of alfalfa fiber eaten.

Different foods have different digestibilities, depending in large part on the extent to which they are impregnated with lignins, tannins, or silica. It follows that herbivores digest some plant foliages more easily and quickly than others. Mule deer (*Odocoelius hemionus;* 100–150 kg) digest most alfalfa fiber in 20–40 hours, but break down far less grass or heavily lignified blueberry foliage in the same amount of time (Fig. 3-12). Because some plant foods are much easier to digest than others, the best feeding strategy for plant-eating animals is to select foods that can be digested within the retention times permitted by the size and activity of their microbial fermentation systems.

Physiological constraints on the digestion of plant fibers force small verte-brates to avoid fiber-rich foods, or use special behavioral and physiological adaptations to overcome the limitations of a small microbial farm. Herbivorous birds usually restrict their diets to fruits, seeds, and buds that are relatively free of fiber. The fiber that is eaten is digested in intestinal ceca. Other small plant-eating vertebrates have special means for handling plant fiber. Rabbits (1 kg) digest only 7–9% of the grass cellulose that they eat for the first time, but they recycle partially digested food by reingesting soft pellets (24–38% protein).

Table 3-5 Mutualisms between microbes and insects

Site	Microbes	Examples
Intracellular		
Abdomen, fat body	Bacteria	Cockroaches (*Blatella*)
Alimentary canal	Rickettsia-like	Cicada (*Magicicada*)
Gut mycetotomes	Eubacteria	Aphids (*Shizaphis*)
Foregut mycetotomes	Bacteria	Grain weevil (*Sitophilus*)
Gut membrane	Bacteria	Wood ant (*Camponotus*)
Extracellular		
Alimentary canal	Flagellates, protozoans and/ or bacteria	Cockroaches (*Blatella*), termites (*Pterotermis*), scarab beetles (*Popillia*), and many others
Gastric ceca	Protozoans and/or bacteria	Milkweed bug (*Oncopeltus*) and other hemipterans
Body cavity	Yeast	Coccids (*Leucanium*)
Fungal Gardens		
Nest	Fungi	Termites (*Nasutitermes*) and leaf-cutter ants (*Atta*)
In wood	Fungi	Bark beetles (*Dendroctonus*)

Source: Data from Jones (1984).

Rabbits also defecate fibrous pellets (9% protein), which are not eaten. Reingesting feces (coprophagy) allows rabbits and rodents to segregate useful protein-rich fractions from fiber, and to reprocess the protein for more-efficient absorption (McBee, 1971). An herbivorous tropical lizard, the iguana (*Iguana iguana;* 1 kg), averages 54% digestibility of partially lignified plant fiber (Troyer, 1984a,b). This surprising accomplishment for a small herbivore is possible in part because a much higher proportion of the lizard's body weight (18.5%) is devoted to digestive and fermentative organs than is the case in small mammals (7.9%). If large vertebrate herbivores are challenged by plant fiber, small plant-eating mammals and birds are under constant pressure to minimize the fraction of indigestible fiber in their diets.

Insects that eat plants are much smaller than their vertebrate counterparts, and consequently cannot rely on a large volume of microbial symbionts for fermentation. Herbivorous insects rely on limited diet selection and a variety of means of maintaining symbiotic microbes to meet the challenge of plant fiber and defensive compounds (Table 3-5). Long guts or elaborate ceca serve as miniature digestive chambers, analogous to vertebrate guts, for nurturing extracellular microbial symbionts (Fig. 3-13). Many insects harbor mutualistic microbes *inside* cells called **mycetocytes** (Fig. 3-14). These endosymbionts play different roles, in different insects, including breakdown of cellulose, fats, proteins, and starches; synthesis of vitamins; and sometimes even nitrogen fixation (Jones, 1984). Mycetocytes may occur in normal insect tissues, or be grouped

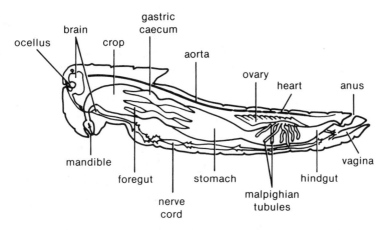

Fig. 3-13 Schematic drawing of a grasshopper gut, showing expanded volumes and ceca required for microflora.

into special organs termed **mycetotomes** found in any of several regions of the digestive tract (Koch, 1967). Cockroaches and some other insects that feed on a variety of plant taxa possess both extracellular and intracellular symbionts, showing a remarkable versatility in the microbial associations that they culture within their bodies.

Some tiny insect herbivores actually cultivate fungal symbionts outside their bodies, thereby freeing their microbial farms from the constraints of body size. Bark beetles (Scolytidae in the order Coleoptera) spread and perhaps cultivate

Fig. 3-14 Mycetocytes in the leaf beetle *Bromius obscurus* (Chrysomelidae). Outpocketings of the anterior (top) and posterior (bottom) midgut are lined with folds of cells containing symbiotic bacteria that assist in digestion. The inset shows these bacteria-containing cells. Other symbionts occur in the Malphigean tubules of this species. From Buchner (1965).

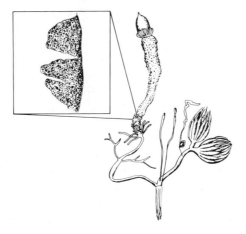

fungi in burrows in wood. The beetles do not eat intact wood, but feed instead on wood that has been partially digested by their symbiotic fungi. Wood-boring grubs of long-horn beetles (Cerambycidae) likewise possess several symbiotic bacteria and yeasts that assist in nitrogen metabolism (Koch, 1967), but these larvae apparently do not have the capacity to digest cellulose without eating fungal extracts. Jerome Kukor and Michael Martin (1986) have even found that larvae of the long-horn *Saperda calcarata,* which normally feed on aspen sap, can become cellulose digesters if they are fed extracts of wood-digesting fungi.

Two groups of social insects, the termites of the family Termitidae (Isoptera) and the leaf-cutting ants of the subfamily Attinii (Hymenoptera) maintain vast fungal gardens that rival or far surpass the volume of gut fluids of the largest ruminants. For instance, leaf-cutter ants *(Atta)* feed on plant sap and the huge fungal colonies that they cultivate with leaf and flower parts in vast underground nests. Not surprisingly, insects that maintain huge fungal gardens forage for a much wider variety of plant species than those dependent on much smaller microbial farms contained within their bodies.

Finally, insect herbivores use behavioral adaptations for meeting the challenge of digesting plant tissues. The most obvious behavioral adjustment is restriction of the diet. With the exception of colonial termites and ants that husband symbionts outside the body, virtually all insects feed on a much smaller range of plant species than do rabbits or buffalo. Many insects feed on only one or a few species of plants in their lives. Foliage-eating adult and larval insects unable to process fiber as efficiently as larger animals must also eat at a faster rate, thereby allowing the digestive system to process easily digestible cell contents and pass most fiber undigested. A caterpillar, for instance, consumes up to 3–4 times its body weight in foliage daily (Slansky and Feeny, 1977).

The necessity of acquiring microbial assistance can be a hurdle for young herbivores. Young mammals inoculate themselves by purposefully or inadvertently ingesting microbes discharged in their mothers' droppings. Young iguanas, hatched in underground nests long after their mothers have returned to the treetops, undertake an elaborate odyssey in search of adults of their species (Troyer, 1982). Once successful, hatchling iguanas eat droppings of adult iguanas before returning to their usual habitat in low second growth. Insects have a variety of means of equipping their offspring with microbial symbionts (Koch, 1967). Most pass bacterial symbionts directly in the egg, or smear them onto the eggshell. Termites feed symbionts directly to their larvae. Social insects appear deliberate in their mode of inoculation (see Price, 1984). Virgin leaf-cutter ant queens carry plant-digesting fungi with them when they leave their home nest to mate, and then carefully nurture the fungi when they start a new nest of their own.

In summary, both vertebrate and invertebrate digestive morphology and physiology are adapted for nurturing microbial symbionts that break down digestibility-reducing polymers in plant tissues. Far from being a "free lunch," plant tissues are fortified with lignins, tannins, and silica that must be either behaviorally avoided or attacked with microbial helpers. In as yet unknown ways, microbial symbionts also degrade plant toxins. Herbivores with large microbial farms, such as large ruminants, large hindgut fermentors, and leaf-cutter ants, can consume a wide array of plant toxins without harm, or can feed them to fungal symbionts that detoxify them with equal thoroughness. Small herbivores do not have this luxury, and must counter plant chemical defenses in other ways.

Chemical Counterattack

Herbivores have more than microbes to protect them against plant secondary compounds. Generalized enzyme systems provide protection against an enormous variety of plant toxins. In addition, some insects seem to have evolved specific detoxification mechanisms for toxins of their regular host plants. Both general and specific protection help herbivores contend with what Lena Brattsten calls "the poisoned platter" of their diets.

Mixed-function oxidases (MFO) are membrane-bound enzymes that detoxify a wide variety of plant and synthetic poisons (Brattsten, 1979a,b). In vertebrates, MFO activity occurs in several organs but is highest in microsomes of the endoplasmic reticulum of liver cells. In insects, the usual sites are fat bodies or the midgut. There are three major characteristics of MFO systems: (1) they catalyze oxidative reactions, resulting in polar products that are easily excreted; (2) they are nonspecific, accepting many chemical substrates; and (3) they are easily induced by exposure to novel toxins. Nonspecificity and induction make MFO systems invaluable to herbivorous animals that eat a variety of plants because they eliminate the need to maintain a wide array of specific protective enzymes.

Detoxification by MFO enzymes occurs in two steps: (1) primary degradation in which a toxic molecule receives a chemical group [e.g., hydroxyl (-OH)] that makes it soluble in water, and (2) conjugation with sugars, amino acids, sulfates, phosphates, or other molecules bound for excretion. Two examples of MFO-catalyzed reactions show the degradation of the synthetic insecticide DDT to kelthane (a reaction that is possible for some genetic strains of flies), and the degradation of the natural insecticide nicotine in tobacco to cotinine (Fig. 3-15). The products, in both cases, are soluble, excretable, and far less toxic than their parent compounds.

MFO activity appears within minutes of exposure to novel chemicals in many insect larvae, and within hours in rats and mice (Brattsten, 1983). An armyworm caterpillar *(Spodoptera eridania)* chews a new leaf and waits for several minutes before starting to eat. Those few minutes induce the MFO activity required to cope with the plant. As hours pass, the caterpillar becomes increasingly efficient at digesting its single food source.

MFO activity varies among species, and among individuals within species. Pharmacologists find that it takes a human 175 minutes to degrade half a dose of the hallucinogen LSD, but that it takes a mouse only 7 minutes to degrade half a dose (Hucker, 1970). Both humans and mice are biochemically prea-

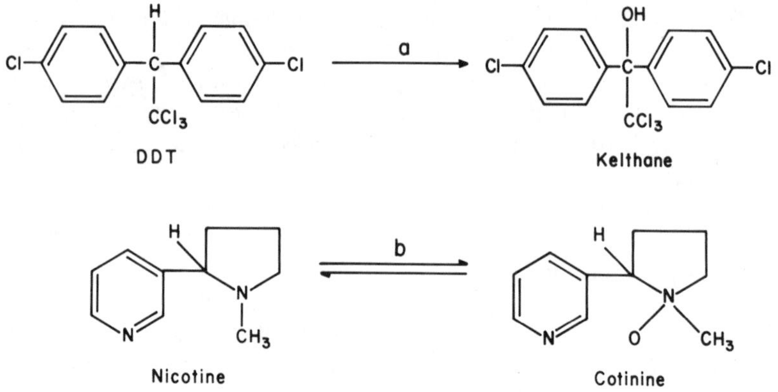

Fig. 3-15 Mixed-function oxidase (MFO) degradations of toxins to excretable forms. (a) Hydrolysis of the synthetic insecticide DDT to the much less toxic kelthane. Some genetic strains of houseflies have a well-developed capacity for this reaction. (b) N-Oxidation of the natural alkaloid nicotine from tobacco into the less toxic cotinine. See Brattsten (1979a, b).

dapted for detoxifying a completely alien poison, but they differ in their ability to do so. Krieger and his colleagues (1971) found that lepidopteran larvae that eat many species of plants have higher MFO activity than those that eat only one or a few. Of 35 species tested, generalists were better adapted than others for degrading novel toxins.

The capacity for insects to develop MFO resistance to novel pesticides is of tremendous concern in agriculture and public health. As Brattsten and her colleagues (1986) point out, insect MFO systems that have evolved for neutralizing natural plant products preadapt many crop pests for developing resistance to synthetic pesticides. MFO systems are evolved, and actively evolving, physiological adaptations of tremendous significance.

Other enzyme systems work simultaneously with the MFO system to detoxify allelochemicals (Brattsten, 1979a). For instance, the enzyme rhodanese converts highly toxic cyanide (CN^-) to the metabolite thiocyanate (SCN^-) which is 200 times less toxic than cyanide. Rhodanese activity is high in such herbivores as cattle and rabbits that come in contact with cyanogenic plants. Rhodanese is one of several **group transfer enzymes** that conjugate a specific toxic entity to an atom (sulfur in the case of rhodanese) or to a larger molecule, thereby making it less toxic than the unconjugated form.

Using Allelochemicals

Some insects use toxins as food. The **nonprotein amino acid** L-canavanine occurs in many species of the bean family (Fabaceae). In constitutes up to 12% of the dry weight of seeds of the tropical bean *Dioclea megacarpa* (Rosenthal, 1977). L-canavanine replaces the amino acid arginine in polypeptide synthesis,

Fig. 3-16   A naive blue jay *(Cyanocitta cristata)* eating a monarch butterfly, and then vomiting. Toxic cardiac glycosides sequestered from milkweeds by monarch caterpillars are retained to the adult stage. The jay will avoid monarchs after this traumatic experience. Photographs by L. Brower. Copyright © 1969 *Scientific American,* reproduced with permission.

thereby producing defective polypeptides that inhibit growth, differentiation, or survival in a wide variety of protozoans, plants, insects, and vertebrates. The toxin protects *Dioclea* from all but one seed-eating insect. The weevil *Caryedes brasiliensis* not only develops entirely within *Dioclea* seeds, it even uses normally toxic L-canavanine as a nitrogen source. Gerald Rosenthal and his colleagues (1982), using [^{15}N]urea, found that the weevil degraded L-canavanine to carbon dioxide (CO_2) and ammonia (NH_3), and then reincorporated the ammonia into 11 of 17 dietary amino acids. Far from avoiding L-canavanine, *Caryedes* lives on it!

Other insects actually incorporate plant toxins into their own protective chemistry (Huheey, 1984). For instance, caterpillar and adult monarch butterflies *(Danaus plexippus)* **sequester,** or store, cardiac glycosides obtained from milkweed species on which the caterpillars feed. These are distasteful to bird predators, which vomit shortly after eating either a monarch caterpillar or adult (Fig. 3-16). The experience is so traumatic that birds avoid future encounters with either the distasteful insect—or even insects that look like it. Avoidance of insects with sequestered allelochemicals promotes the evolution of bright warning coloration among distasteful insects, in some cases even leading to coevolved mimicry complexes of noxious species (see Chapter 8).

Choice and Avoidance

Animals that eat many plant species in a lifetime (**polyphagous** herbivores) face a different world from those that eat only a few plant species (**oligophagous** herbivores), or even just one species (**monophagous** herbivores). Polyphagous animals must use microbial symbionts and extensive physiological defenses to counter a wide variety of plant toxins and fibers. Animals that feed on fewer species must sense and behaviorally avoid plants that their digestive and fermentative systems cannot handle.

Most vertebrate herbivores, snails, slugs, and many insects are polyphagous. Few if any mammals eat only one species of plant (Freeland and Janzen, 1974). Even a specialist like the Australian koala *Phascolarctos cinereus* eats many species of a very large genus, *Eucalyptus*. Mammals frequently reject some foliage on the basis of smell, but use sight, smell, taste, and memory to select up to dozens of food plants from among hundreds in their environments. Much of this is trial and error. Individuals of foliage-eating monkeys (e.g., *Allouatta, Colobus),* deer *(Odocoelius),* and rabbits *(Sylvilagus)* confine daily diets to substantial quantities of only a few plant species, thereby avoiding the danger of overwhelming both their MFO systems and gut flora with novel toxins. At the same time, these animals taste small quantities of new foods daily, thereby inducing specific MFO activity or stimulating a change in gut flora (e.g., Milton, 1978). Freeland and Janzen found that captive tropical forest-rats *(Tylomys)* nibble seeds of many species rather than settle for one favorite kind. For such polyphagous creatures, foraging requires far more than assimilating energy and nutrients. These animals must also cultivate their microbial cultures and nurture a dynamic digestive chemistry.

Polyphagous insects normally feed on only one plant species each day. Because a polyphagous insect that weighs only 0.1 to 1.0 g need only cope with a limited range of plant secondary compounds each day, it relies primarily on MFO induction in its own tissues (Brattsten, 1979b, 1983). Large polyphagous vertebrates meet far more diverse daily chemical challenges with extensive use of microbial detoxification along with MFO systems in their own tissues.

The primary challenge for oligophagous and monophagous insects is finding food, not detoxifying a variety of foods. In general, insect sensory abilities match their needs for distinguishing toxic and edible host plants in their environment (Rausher, 1983; Miller and Strickler, 1984). Highly specific chemosensory abilities allow insects with the most limited diets to sense much of the world as simply "wrong chemistry." For instance, the cabbage moth *(Mamestra brassicae)* possesses chemoreceptor cells that respond only to glucosinolates from its host plants in the mustard family (Cruciferae). These receptors do not detect salts, sugars, or tannins, but respond strongly to a range of glucosinolates and related compounds that most insects find repellent (Wieczorek, 1976). Cabbage moth caterpillars sense light, touch, and a few chemicals, but they simply fail to detect much of their environment. On the other hand, insects with broader diets may select hosts on the basis of relative abundances (Singer, 1982). For instance, the butterfly *Euphydryas editha* will lay its eggs on host plants that are usually not preferred, if such species are much more common than hosts that are usually favored. This kind of discrimination requires broader sensory capabilities and a higher degree of behavorial flexibility than is necessary for cabbage moths.

Overview

Overcoming complex plant defenses requires complex feeding adaptations. Teeth or mandibles, gut microbes, and MFO systems make it unnecessary for herbivore species to evolve independent means of countering each specific plant defense; however, animals do differ in their ability to use different plant foods. Few very small (<1 g) species are capable of feeding on more than a few species or, at most, a few families of plants. This occurs because plant defenses have both distinctive and synergistic effects on herbivore food choice.

Each plant species contains several digestibility-reducing and toxic compounds that defend it against different herbivore enemies. Distinctive effects of different toxins can be critical. Even if a weevil, for instance, can detoxify *all but one* allelochemical of a seed, the seed is still safe from that insect enemy. A diverse array of secondary compounds is important because of the individual toxicity of those chemicals.

The combined effects of several allelochemicals *are often greater* than the sum of their separate effects. Enzyme defenses against toxins take energy, and are usually induced to handle particular allelochemicals rather than all possible toxins at once. A battery of different allelochemicals in an especially heavily defended plant may overwhelm gut microbes and enzyme systems far better than high concentrations of most single toxins. A diverse allelochemical load may have a dosage-dependent effect on herbivores that could detoxify each separate allelochemical.

Plants gain from both the individual and combined effects of digestibility reducers and toxins in their tissues. Their ability to survive in a world of plant-eating animals depends on their ability to avoid herbivores in space and time, and on their capacity to assimilate carbon and nutrients necessary to synthesize those defenses. Animals need more than morphological adaptations and mechanical force to overcome plant defenses. Herbivores either employ an enormous volume of plant-processing microbial symbionts, use sensory adaptations and foraging behaviors that allow selection of foods that less efficient microbial and MFO systems can accommodate, or employ some intermediate combination of foraging selectivity and microbial and enzyme assistance.

SUMMARY

Plants have evolved an impressive arsenal of mechanical and chemical defenses against pathogens and herbivores. Cellulose and hemicellulose support plant tissues with fibers that are, when fortified with lignins, nearly indigestible to most animals. These polymers in combination with tannins or silica dramatically reduce the food value of many mature tissues to all but the best-equipped herbivores. Most land plants are also fortified with secondary compounds that are, in small quantities, toxic to most herbivores and pathogens. Toxins may be a permanent part of a plant defensive system, or be induced when needed.

A parallel history has given herbivorous animals morphological, biochemical, and behavioral tools for coping with the "poisoned platter." Teeth, mandibles, and gizzards grind up plant tissues, and elaborations of the digestive tract house mutualistic bacteria and protozoans that digest cellulose and hemicellulose and detoxify allelochemicals. Herbivores also induce (MFO systems) or maintain (group transfer enzymes) a variety of enzymes that detoxify allelochemicals by converting them to forms that can be harmlessly excreted, or even used for animal nutrition. Sensory capabilities and complex foraging behaviors allow animals that eat many plant species to evaluate foods and select among them.

The ecological struggle between plants and herbivores is dynamic. The eaters and the eaten vary in distribution and abundance, and have the capacity to adapt to each other. Rapidly expanding interest in both the mechanics of herbivory and plant defense and the consequences of these interactions for natural and agricultural communities make the ecology of herbivory one of the most exciting areas of field and experimental science today.

STUDY QUESTIONS

1. A friend decides that he wants to avoid toxic food additives by eating only "natural" uncooked vegetables and wild herbs. How would you advise him?

2. A physician in equatorial Africa once tried to keep fruit-eating bats from demolishing fruit crops around his yard. He injected fruits with a dose of strychnine sufficient to kill 35–40 humans, each. The bats ate the fruits without ill effects. How might you explain the doctor's bad luck, and the bats' good fortune?

3. Given what you know about MFO systems, how might you expect polyphagous and monophagous insects to respond to a new insecticide?

4. Insects frequently evolve a resistance to novel synthetic pesticides, but have in most cases failed to overcome host defenses based on ecdysone analogues. First, why are ecdysone defenses difficult to overcome? Second, should commercial insecticides employ ecdysones? Why or why not?

5. If a diverse battery of chemical defenses is the best way for plants to ensure safety from herbivores, what prevents all plants from evolving total toxicity to all herbivores?

SUGGESTED READING

John B. Harborne (1977) provides an excellent introduction to chemical ecology in *Introduction to Ecological Biochemistry*. Gerald Rosenthal and Daniel Janzen (1979) cover the mechanics of plant defense and herbivory in an edited volume entitled *Herbivores: Their Interaction with Secondary Plant Metabolites*. *Chemical Ecology of Insects*, edited by William Bell and Ring Carde (1984) gives more attention to sensory capabilities and behavior. Peter Van Soest (1982) and Charles Robbins (1983) provide comprehensive discussions of vertebrate herbivory in *Nutritional Ecology of the Ruminant* and *Wildlife Feeding and Nutrition*, respectively. Also see the Suggested Reading for Chapters 4 and 5.

4

Ecology of Herbivory

Herbivores are devastating when they breach plant defenses. Oak forests may be completely defoliated by moth caterpillars (Fig. 4-1). Less obvious dramas underneath healthy adult trees of many species doom virtually all undisseminated seeds to devastating mortality from insects, rodents, and disease. Even subtler dramas exist. Leaf-eating insects depress seed production and seed viability in *Piper* shrubs for at least 2 years after defoliation ends (Fig. 4-2), indicating a silent struggle between plants and their enemies that would pass unnoticed without careful experimentation. Whether animals kill plants outright or cause less obvious havoc by chewing buds, burrowing through leaves, or by

Fig. 4-1 Defoliation of an oak *(Quercus robur)* forest by winter moth *(Operophtera brumata)* caterpillars. Bare branches would ordinarily be hidden in the spring foliage. From Feeny (1976).

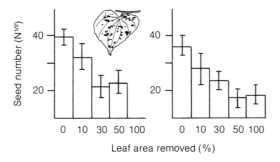

Fig. 4-2 Effects of leaf area removal from *Piper arieianum* shrubs on seed production 1 year (left) and 2 years (right) after a single defoliation in Costa Rica. Partial defoliation has lasting effects on both seed number and viability. This experiment simulates natural weevil (*Perdinetus* sp.; *Ambetes* sp.) herbivory. After Marquis (1984).

boring into bark or seeds, herbivory often tips the balance of natural selection against poorly defended trees, shrubs, or herbs.

Yet green plants surround us. Somehow, millions of herbs, shrubs, and trees outgrow, outlast, or outdefend their herbivore enemies. One of the principal challenges of modern ecology is to understand how plants escape from herbivores in time and space. Why are some species successful when others are not? Why do some individuals survive unscathed while their neighbors are chewed to ribbons? Why do trees escape defoliation in some years, but are denuded by millions of ravenous caterpillars in others?

This chapter probes the patterns of plant defense and animal use of plants. Plant defense theories provide the organizing principles. The ease with which plants are located by herbivores provides one theoretical perspective (apparency theory); the ease with which plants mobilize resources to defend themselves provides another (resource availability theory). Both are grounded in the idea that plants evolve defensive capabilities, that are countered by adaptive evolution of animals that eat them (classical theory). Selected examples then ascertain how well pieces of empirical discovery fit the puzzles of alternative theoretical perspectives. Future chapters (Chapters 9 and 10) probe the effects of these patterns on the distribution and abundance of plant and animal species in ancient and modern communities.

CLASSICAL PLANT DEFENSE THEORY

Plant defense theory has had a long evolution. Dethier (1954) and Fraenkel (1959) noticed that different species of insects react differently to toxic plant chemicals. Dethier and, later, Ehrlich and Raven (1964) elaborated what is now a classical theory of **biochemical coevolution** between insects and plants. The central idea is that plant species evolve secondary compounds in response to attacks by insects, while insects meet the challenge by evolving new detoxification systems. In adapting to the secondary compounds of particular plant families, insects lose the ability to detoxify allelochemicals of unrelated plants. Plant families eventually acquire a complex of defenses that exclude all but a fauna of specialist herbivores. The classical theory predicts that related taxa of insects become locked into a chemical arms race with related taxa of plants.

The classical theory of biochemical coevolution has been the catalyst for an immense amount of research that addresses the deceptively simple question, "Why do animals eat some plants, and not others?" Supporting evidence for the theory comes from herbaceous plants with simple toxins. For instance, Berenbaum (1983) finds that caterpillars of some swallowtail butterfly species eat members of the parsnip family that are highly toxic to other insects. Less toxic parsnip species are edible for many insects. Other examples are less compatible with the classical theory. Most vertebrate and many insect herbivores have broader feeding preferences than the theory predicts. More general theories must be devised to explain why plant-eating animals differ so much in dietary specialization, and why plants themselves differ so widely in vulnerabilities to enemies as diverse as aphids and elephants.

PLANT APPARENCY: A THEORY OF HIDE AND SEEK

The theory of **plant apparency** holds that plants easily found by herbivores evolve different kinds of chemical defenses from those that are difficult for animals to locate. Developed by Feeny (1976) and Rhoades and Cates (1976), this theory proposes a game of hide-and-seek between the eaters and the eaten. Plants that are easy for herbivores to locate should invest heavily in quantitative digestibility reducers that provide generalized protection against all herbivores (Table 4-1). Plants that are difficult to locate should rely on escape in space and time, and on small amounts of qualitative toxins that are effective against all but specialist herbivores.

Apparency theory is a theory of plant defense from the animal—particularly insect—perspective. **Apparent** plants are those that are easy for herbivores to find, while **unapparent** plants are difficult for plant-eating animals to find. Trees and shrubs of the climax forest and perennial grasses of prairies and savannas are apparent plants that live dozens to hundreds of years and cannot avoid eventual colonization by insects. Such species are thought to defend themselves with tough nutrition-poor leaves and with indigestible lignins, tannins, or silica that further reduce nutritional value with a dosage-dependent quantitative effect on all herbivores.

Short-lived herbaceous plants of early successional stages are unapparent because they are less likely to be discovered by insects during the few months or years that they grow in a forest clearing or in a recently disturbed prairie. Not needing generalized defenses, unapparent plants should rely on simple toxins with a qualitative effect on herbivore metabolism. Such toxins as alkaloids,

Table 4-1 Ecological correlates of plant defenses, according to apparency theory

	Qualitative Defenses	Quantitative Defenses
Examples	Alkaloids, cyanogens, glucosinolates, nonprotein amino acids, terpenes	Cellulose, hemicullulose, lignins, tannins, silica[a]
Properties	Small toxic molecules	Complex polymers or crystals that reduce digestibility
Amounts	Low (<2% dry weight of plant tissues)	High (>85% dry weight of plant tissues)
Distribution within plant	New or especially valuable leaves, buds, or ripening fruits (unapparent or unpredictable tissues for herbivores)	Permanent woody tissues or mature leaves (apparent or predictable tissues for herbivores)
Distribution among plants	Rare, short-lived, herbaceous, and/or early successional species (unapparent or unpredictable species for herbivores)	Common, long-lived, woody, and/or late successional species (apparent or predictable species for herbivores)
Phylogeny	Most common in advanced angiosperms	Prevalent in ancient ferns, gymnosperms, angiosperms

[a]Cellulose and hemicellulose provide support and reduce digestibility, particularly to omnivores that eat plant parts (e.g., foxes).

glucosinolates, and terpenes are poisonous to most herbivores, but are easily detoxified by specialists that have evolved enzyme defenses appropriate to their normal food plants. Because unapparent plants are short-lived and scarce, however, few insects have the opportunity to find them and evolve means of detoxifying their secondary compounds. Toxins at low concentrations (often <1% dry weight) suffice.

The principle of apparency can also be applied to parts of plants. For instance, Doyle McKey (1979) argues that plants should allocate resources within their tissues in a manner that most effectively defends them against herbivores. Permanent quantitative defenses are expected in permanent woody tissues or mature leaves, while inexpensive qualitative defenses should characterize ephemeral but especially valuable tissues or organs, such as new leaves or seeds.

Oaks and Mustards

The theory of plant apparency owes much to the insights of one scientist who studied insect herbivory on two very different kinds of plants. Paul Feeny's (1970, 1976) comprehensive studies of oaks in England and mustards in central New York still provide the clearest contrasts for apparency theory.

The common oak *(Quercus robur)* of England and continental Europe typi-

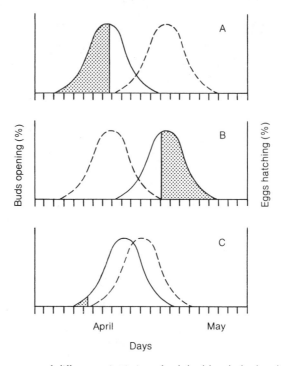

Fig. 4-3 Consequences of differences in timing of oak bud break (broken line) and winter moth egg hatch (solid line). (A) Egg hatch precedes bud opening; early-hatching larvae (stippled) starve. (B) Bud opening precedes egg hatch; late-hatching larvae (stippled) starve. (C) Egg hatch broadly coincides with bud opening; defoliation occurs. After Feeny (1976).

fies an apparent plant species. Individual oak trees live for decades to centuries, and possess dense wood and tough leathery leaves that are difficult for herbivores to digest. Oak-eating insects usually grow slowly, occur in low population densities, and rarely destroy much foliage. It was the *occasional* devastating defoliations of English oaks (Fig. 4-1) by larvae of the winter moth *(Operophtera brumata)* that attracted Feeny's attention.

Timing influences the uncertain relationship between oaks and winter moths. Wingless female winter moths emerge from pupae in November or December, climb up the nearest vertical object, often an oak in the species-poor forests of England, and lay their eggs. The timing of egg hatch and bud break determines caterpillar success (Fig. 4-3). If the eggs hatch before the oak buds break, no food is available and the caterpillars starve. If the eggs hatch after most of the buds break, the tiny caterpillars starve because they cannot eat toughening, tannin-rich leaves. Only when egg hatch and bud break overlap broadly do most larvae find suitable food. The oak has a brief window of vulnerability when good fortune for the winter moth spells havoc for the tree.

The relationship between oak leaf maturation and herbivory illustrates why late-hatching moth larvae are doomed (Feeny, 1970). Larvae of winter moths

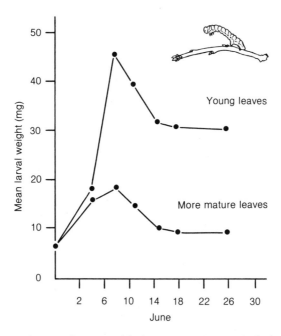

Fig. 4-4 Winter moth caterpillar size and leaf age. Larvae increase in fresh weight until just before pupation, at which time they decline to the final pupal weight. Maximum weights of larvae raised on protein-rich and tannin-poor young leaves are three times higher than those of larvae raised on protein-poor and tannin-rich mature leaves. After Feeny (1970).

raised on oak leaves produced early in the season are up to three times heavier than those raised on leaves produced only two or three weeks later (Fig. 4-4). Analysis of leaf composition shows that early leaves have high levels of protein and low levels of tannins. As the season progresses, protein content drops from 40 to 15% dry weight, while tannin content climbs steadily from <1 to nearly 6% dry weight. Feeny found that as little as 1% dry weight of tannin added to artificial diets stunts larval moth growth and reduces fecundity of those that do survive to adulthood. A combination of tannin production and increasing leaf toughness protects oaks during years that winter moths hatch after oaks leaf out.

Early successional herbs that live one or a very few years provide a contrast to the apparent oaks. The mustard family (Cruciferae) includes such common garden vegetables as kale and cabbage *(Brassica)* and a variety of wild herbs (e.g., *Barbarea, Brassica, Hesperis, Lepidium, Thlaspi*). Most mustards (members of the Cruciferae) invade open ground and are quickly displaced by grasses or woody plants during normal ecological succession.

The contrast between oak and mustard chemistry is striking. The complex phenols and lignified cellulose that are so important to oaks are absent from mustards. The mustards contain small quantities (often <<1% dry weight) of glucosinolates that are absent from oaks. As many as nine different gluco-

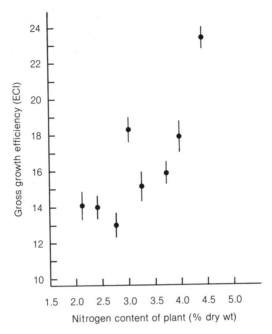

Fig. 4-5 Relationship between gross growth efficiency (ECI) of fifth instar cabbage butterfly *(Pieris rapae)* caterpillars and nitrogen content of cultivated and wild mustards. As specialists on members of the family Cruciferae, cabbage butterfly growth closely reflects the nitrogen content of host leaves, despite wide differences in glucosinolate content. Each point (± 1 SE) represents determinations for one species or cultivar. After Slansky and Feeny (1977).

sinolates are present in cultivated cabbages (varieties of *Brassica oleracea*), and even more occur in wild species (Slansky and Feeny, 1977).

Glucosinolates vary widely in their toxic effects. Some of these compounds and the mustard oils derived from them are poisonous to most insects. Black swallowtail *(Papilio polyxenes)* larvae avoid mustards in the field, but in the laboratory will eat celery (an acceptable food plant) laced with the glucosinolate sinigrin. Concentrations as low as 0.001% sinigrin stunt growth and reduce fertility, and concentrations of as little as 0.1% kill the caterpillars (Erickson and Feeny, 1974). In contrast, Slansky and Feeny found that glucosinolates had little or no effect on larval growth or survival of a mustard specialist, the cabbage butterfly *(Pieris rapae)*. The nitrogen content of wild and cultivated mustards, not glucosinolate content, determines the growth efficiency (ECI in Appendix II) of cabbage butterfly caterpillars (Fig. 4-5). If these specialists find a mustard patch, the insects prosper at the plants' expense.

Plant apparency theory emphasizes extreme defensive strategies. Oak survival is thought to depend on tough leaves and digestibility-reducing compounds that deter, weaken, and sometimes kill a wide variety of herbivores that collect on the trees during their long lives. These apparent plants use generalized defenses against generalized herbivores. Mustards rely on a few simple compounds that are distasteful or lethal to all but the few herbivores specialized to eat them. These unapparent plants escape specialists because of their scattered distributions in ephemeral successional habitats.

Limits of Apparency Theory

The dramatic contrast between oak and mustard life histories shows the theory of plant apparency in its best light. It would be optimistic to expect any theory to fully explain patterns of defense among tens of thousands of forbs, grasses, shrubs, and trees. Are oaks and mustards representative of all plants? Do most trees actually rely on digestibility-reducing compounds, and most herbaceous plants on toxins? Plant-eating animals range in size and digestive physiology from aphids (0.005 g) to elephants (5,000,000 g). Can apparency theory predict their foraging behavior and food preferences? These are controversial questions. A wealth of phenomena lies between a cabbage and an oak, or an aphid and an elephant.

PATTERNS OF DEFENSE IN PLANTS

How well do predictions from apparency theory actually fit patterns of plant defense? Some obvious clarifications are needed at the outset. For instance, individual grass stems live only a season, but the underground plant that produces them may live for hundreds or thousands of years. These are apparent plants, even though they are not woody. Consistent with apparency theory,

Table 4-2 Defensive chemical types known from woody (apparent) and herbaceous (unapparent) plant types

Chemical Types	Woody (% of families)	Herbaceous (% of families)
Digestibility Reducers: Quantitative Defenses		
Phenolics[a]	85	38
Toxins: Qualitative Defenses		
Quinones	38	69
Saponins	8	46
Alkaloids	23	77
Cyanogenic glycosides	23	38
Coumarin glycosides	23	38
Acetylenes	0	23
Sulfur compounds	0	23

Source: After Futuyma (1976).

[a]Phenolics include digestibility-reducing tannins.

grasses are fortified with silica, which serves the same digestibility-reducing role as lignins in woody plants.

A second clarification concerns the chemical distinctions between woody and herbaceous dicots. Douglas Futuyma (1976) used the published literature to test the idea that apparent species have quantitative defenses, while unapparent species rely on qualitative defenses. The distinction exists, but it is not absolute (Table 4-2). Some woody species have qualitative defenses in addition to digestibility-reducing lignins and phenols, and some herbs have high levels of phenols in addition to toxins. This lack of a clear allelochemical distinction between the two kinds of plants raises questions that must be explored. A series of explicit hypotheses concerning patterns of plant defense, and animal use of plants for food, must be spelled out and tested in a variety of field and laboratory situations.

Apparency and Defense: Persistent and Pioneer Trees

Can predictions from apparency theory stand up to tests in plants other than the oaks and mustards in which they were originally framed? The tropical rainforest of Barro Colorado Island, Panama, contains at least 400 species of trees in an area of only 15 km^2 (~9 square miles). Many of these tree species are **persistents** of the rainforest. They establish as seedlings in the understory and grow in the shade for most or all of their lives. Others are **pioneers** that establish and grow only in light gaps opened by treefalls. All of the common persistents and pioneers are apparent trees when compared with a mustard in an old field, but they differ substantially from each other in their life histories (Table 4-3). Persistents grow more slowly than pioneers and may live for one or two centuries. Pioneers grow quickly and die within 15 to 30 years. Do patterns of defense and herbivory in tropical persistents and pioneers parallel those of temperate oaks and mustards, as defense theory would predict?

Phyllis Coley (1983) tested apparency theory in the forest of Barro Colorado Island by comparing leaf defenses and grazing damage on persistent and pioneer species. As expected, mature leaves of persistents were tougher, had less water, and were better defended with tannins, lignins, and cellulose than mature leaves of pioneers (Table 4-4). Quite unexpectedly, young leaves of both persistents and pioneers had up to three times more tannins than mature leaves. Recall that herbaceous mustards lacked tannins, and oaks lacked tannins in young leaves. Patterns of defense only partly conformed to the oak and mustard dichotomy.

Patterns of herbivory also both met and contradicted predictions from apparency theory. As expected, mature leaves of persistents were grazed much less (0.04%/day) than mature leaves of pioneers (0.24%/day). Ironically, pioneers did not escape herbivory by virtue of having high tannin contents in short-lived young leaves. Young leaves of both persistents and pioneers are much more heavily grazed than mature leaves (0.83 and 0.97% per day, respectively). Nitrogen-rich young leaves did not escape herbivory by virtue of their short

Table 4-3 Annual growth rates for saplings of pioneer and persistent tree species in lowland Panama

	Pioneers (20 species)	Persistents (21 species)
Growth in height		
Mean (cm/year)	96	37**
Maximum (cm/year)	155	86**
Variance	3365	855**
Leaf production		
Number of leaves/year	46	38
Leaf area (cm²/year)	6026	3700*

Source: After Coley (1983).

$*p < .05$; $**p < .01$ with the Mann Whitney U Test, using 142 pioneers and 159 persistents.

exposure to herbivores, nor their unexpected fortification with tannins. Neither of these results were predicted by apparency theory.

One implication of Coley's study is that tough lignified foliage is far more important as a generalized defense than tannins. Whatever role tannins play in young leaves, they do not deter *some* herbivores from doing extensive damage. More tannin-rich young leaves are eaten than lignified old leaves. The digestibility-reducing role of tannins is not as generalized as apparency theory predicts (Bernays, 1981). It is quite likely that these leaves would be more heavily damaged without *any* tannins, but it is clear that some herbivores have counteradapted to them, much as herbivores often counteradapt to simple toxins.

Table 4-4 Defensive characteristics of young and mature leaves of pioneer and persistent canopy tree species of lowland Panama (mean values for each character are given)

	Pioneer (22 species)		Persistent (24 species)	
	Young	Mature	Young	Mature
Chemical				
(% dry weight)				
Total phenols	13	8	19	10
Tannin	3	2	10	5
Lignin	10	10	12	12
Cellulose	17	18	20	23
Physical				
Toughness (N)[a]	2	4	2	6
Hairs (no./mm²)[b]	6	5	1	1
Nutritional				
Water (%)	74	70	76	62
Nitrogen (% dry weight)	3	2	3	2

Source: Data from Coley (1983).

[a] Newtons required to punch a 5-mm-diameter rod through a leaf.

[b] Undersides of leaves.

This tropical study broadens our understanding of apparency, while exposing some deficiencies of apparency theory. Persistent species have much tougher foliage than pioneers, as predicted by apparency theory. This is true even though persistents are apparent only 3 or 4 times as long as pioneers. By comparison, English oaks live 50 to 100 times as long as mustards. However, young leaves on tropical trees have sufficient tannin to kill winter moths. Perhaps, from the perspective of insects searching for palatable foliage, young leaves of English oaks that flush over a very short time in the temperate Spring are much less apparent than those of tropical trees that leaf out in a world full of plant-eating animals. Pioneers and persistents alike must protect their tender young foliage much more thoroughly in a world of nondormant herbivores. The lesson here is that apparency is difficult to define in ways that can apply to widely different plants in very different communities.

Apparency and Defense: Parsnips and Relatives

Degrees of apparency also exist in herbaceous dicots. May Berenbaum (1981, 1983) compared the distributions of insect herbivores on plants containing linear furanocoumarins, more-derived angular furanocoumarins, and the precursors to each, hydroxycoumarins (Fig. 4-6). Hydroxycoumarins are widespread and not particularly toxic to insects. In the parsnip family (Umbelliferae), gen-

Fig. 4-6 Furanocoumarin skeletons and their hydroxycoumarin precursors in parsnips and other plants. Marginally toxic hydroxycoumarins occur in hundreds of genera. Linear furanocoumarins are restricted to 35 plant genera and deter all but specialized insect herbivores. Angular furanocoumarins, which are activated by ultraviolet light in sunshine, occur in 11 of the genera containing linear furanocoumarins and are lethal to almost all insects. After Berenbaum (1983).

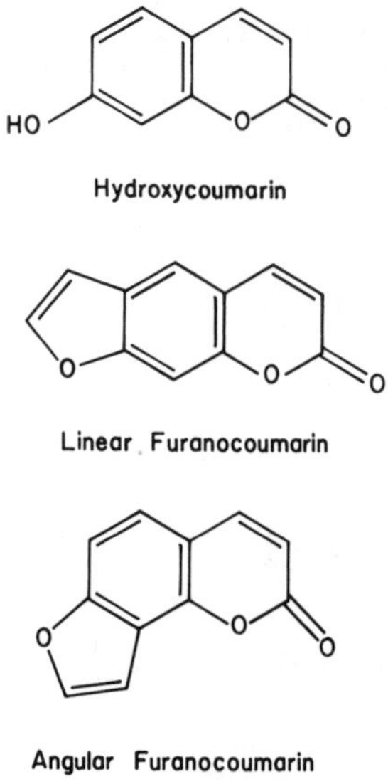

Hydroxycoumarin

Linear Furanocoumarin

Angular Furanocoumarin

era with linear furanocoumarins have diverse herbivore assemblages, while those with the more-derived angular forms are eaten by only a few specialists like moths of the genera *Agonopterix* and *Depressaria*. Both linear and angular furanocoumarins are activated by ultraviolet light in sunshine. Moths feeding on plants containing these compounds avoid eating active toxins by rolling up leaves and feeding on shaded tissues.

As expected by apparency theory, umbellifers that occupy a wide variety of habitats are protected by the most generalized hydroxycoumarins and linear furanocoumarins. These are also the plants with the widest variety of herbivores. The most lethal toxins, the angular furanocoumarins, are found in early successional plants. Like persistent and pioneer trees, herbaceous umbellifers show clear differences among themselves in both apparency in insects and in defenses against their enemies.

Changes in Defense with Time

Plant defenses change with time. Leaves toughen as they mature, and sometimes potent toxins in young shoots are replaced by lignified or silicified tissues in older leaves and twigs. Other plants respond directly to herbivore damage, either by producing toxic chemicals or by compensatory growth or reproduction.

Maturation

Virtually all plants and plant shoots become less palatable as they mature. On the Serengeti Plains in Tanzania, East Africa, grasses sprout from underground

Fig. 4-7 Relationship between the total yield of grass biomass, yield of digestible matter, and stage of plant growth. As grasses mature, a smaller and smaller proportion of tissues is digestible to herbivores. Beyond a point that is different for each species, nutrient content actually decreases as foliage becomes more and more silicified. Lignification during maturation of many broad-leaved plants similarly reduces digestibility. After Van Soest (1982).

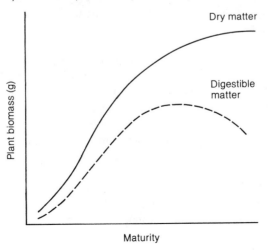

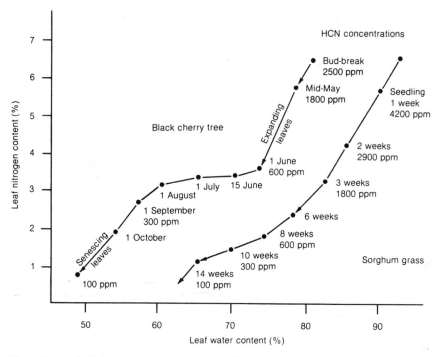

Fig. 4-8 Seasonal changes in nutrient and cyanide content in a tree (black cherry) and a grass (sorghum). Nitrogen and water content are highest early in the season, when soft immature leaves or seedlings are well protected with cyanogenic glycosides. As the season progresses, nitrogen and water content decrease as leaves toughen and become less nutritious. This decrease in nutritional quality is accompanied by a dramatic decrease in protective toxins. After Scriber (1984).

buds shortly after rains begin. At this time they are nitrogen-rich, low in fiber, and are palatable to most ungulates. As they mature, the yield in weight of grasses increases, but digestibility declines as they become fortified with silica (Fig. 4-7). All but the largest herbivores must then select carefully, or starve in a sea of grass.

Sometimes plants trade qualitative defenses for protection by digestibility reducers as the season progresses (Scriber, 1984). Nitrogen-rich young foliage of sorghum grass and black cherry trees would be choice forage if they were not well stocked with cyanogenic glycosides. As the season progresses, both cyanogen and nitrogen contents decline as lignification or silicification proceeds (Fig. 4-8). As with the tropical plants studied by Coley, these temperate species protect particularly nitrogen-rich foliage with exceptional chemical measures.

Chemical Responses to Herbivory

The response of a plant to attack from herbivores or pathogens depends in part on its condition, and in part on its inherited ability to reallocate resources. If serious stress diverts energy from chemical defense, plants or even parts of plants weakened by shade, disease, poor soil, or recurrent defoliation may be vulnerable to herbivory. Weakened plants are especially susceptible to devastating attacks by insects and pathogens, including many that do not normally harm the host species (Rhoades, 1983, 1985). If an entire population is under moisture, heat, or nutrient stress, massive attacks by sawflies, moth larvae, bark beetles, needle miners, scale insects, or aphids can defoliate or kill enormous numbers of trees. More commonly, very local stess *within* an individual tree or shrub makes some branches more susceptible to herbivory than others. This occurs because branches photosynthesize many of their own energy reserves that are used for defense as well as for growth and maintenance.

Plants healthy enough to respond facultatively to insect or microbe attack sometimes fight back directly with increased toxin production. Chemists have found that proteinase inhibitors are induced by potatoes and tomatoes as a direct response to herbivore damage (Ryan, 1983), but ecologists are now finding evidence that induced responses are far more widespread in plants than had previously been suspected.

Mountain birch *(Betula pubescens)* is a European relative of several common

Fig. 4-9 Survival and fecundity of moth larvae *(Epirrita autumnata)* on leaves of birch trees *(Betula pubescens* spp. *tortuosa)*. Either leaf damage or application of caterpillar droppings (frass) increases the resistance of the trees, and consequently decreases correlates of moth fitness. Bars indicate standard deviations. After Haukioja et al. (1985b).

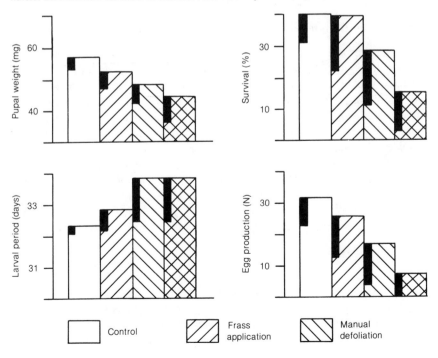

American trees. A team headed by Erkki Haukioja has conducted a long-term investigation of wound-induced defenses of this species in Finnish Lapland (reviewed by Haukioja, 1980). When insects or an investigator damaged mountain birch leaves, neighboring leaves on the same plant became less suitable for caterpillar food. For instance, Haukioja et. al. (1985b) found that wound-induced resistance reduced the growth rate and final pupal weight of an important geometrid (inchworm) defoliator *(Epirrita autumnata)*. A consequence was that plants damaged early in the season had the ability to protect themselves from later attack. Even more startling was the discovery that exposure to insect droppings (frass) was as effective as damage itself in reducing palatability to caterpillars the following year (Fig. 4-9). Mountain birch somehow responded to either direct damage or to a chemical cue that direct damage was likely during the next generation of moths.

The mechanism of induced resistance in mountain birch is under active investigation. Wounded plants have higher total phenol content than others, suggesting that injured plants invest in defensive digestibility reducers rather than growth (Haukioja et al., 1985a). Furthermore, plants with high phenol contents have fewer insect herbivores than others. These results are especially interesting because birch phenols are apparently *not* permanent fixtures of leaves, as apparency theory would predict. They are induced much as a simple toxin might be induced. In fact, it is not clear whether phenols are the actual defensive chemicals responsible for reduced palatability of mountain birch to inchworms, or whether phenols are merely indicators of proteinase inhibitors or other induced defensive chemicals.

Reproductive Compensation

Plants may respond to herbivory with **reproductive compensation,** or excess production of seeds. Stephen Hendrix (1984) has found that the furanocoumarin defenses of cow parsnips *(Heracleum lanatum)* are not sufficient to prevent a tiny caterpillar *(Depressaria pastinacella)* from eating their flowers. The plants compensate for a loss of flowers early in the season by producing a huge proportion of female flowers late in the season. Undamaged plants produce a mix of male and female flowers. Damaged plants reproduce through seeds rather than through less dependable pollen dissemination. Reproductive compensation does not repel or poison herbivores, but it does illustrate an additional way that plants respond actively to the challenges of mandibles and teeth.

ANIMAL FEEDING SPECIFICITY

Diet breadth determines the challenges to a gut microflora and MFO system. A grasshopper lives a month or two and eats dozens of individual plants during its lifetime. An elephant may live 70 years and eat hundreds of thousands of plants or plant parts. Animals that face the challenges of many different foods during a lifetime occupy what Richard Levins (1968) once called **fine-grained** environments. In the course of a day or year, they face innumerable ecological circumstances that force them to select or reject food, and live with the biochemical consequences. In contrast, a weevil grub may spend its entire life as a destructive herbivore in one seed, a caterpillar may hatch, feed, and pupate on one tree, and a gall aphid is likely to be born and die on a single leaf. Life for such creatures is **coarse-grained,** meaning that most of all of an individual existence is encompassed by one or, at most, a very few feeding places. Insects on a single host make few choices, and must adjust to only a narrow range of plant toxins. Perhaps no distinction is more fundamental in the ecology of herbivory than that between generalists living in a fine-grained world and specialists adjusting to a coarse-grained existence.

Eating Many Plants

Animals that must select among dozens to millions of plants during their lives include all foliage or seed-eating mammals and birds, algae-eating fish, and innumerable slugs, beetles, grasshoppers, and seed-eating and leaf-cutting ants, to name but a few. These are clearly polyphagous herbivores as both individuals and species. Is apparency theory relevant to diet selection by these grazers, browsers, nibblers, and harvesters?

The Serengeti Plains of East Africa offer an excellent opportunity to compare the foraging tactics of large herbivores (Jarman and Sinclair, 1979). How can all of these animals subsist in what appears to be the same simple grassland niche? Do they partition food species by fiber content, or are different mammal species biochemically adapted to different plants?

A survey of the feeding ecology of three Serengeti herbivores shows that the animals partition food plants by fiber content (Fig. 4-10). Impala *(Aepyceros melampus)* are medium-sized antelope (40–55 kg) that are highly selective grazers and browsers. Small rumens and high metabolic needs force impala to choose easily digestible foliage and protein-rich fruits and seeds whenever they are available. Impala select among plants, perhaps reflecting the need to avoid toxic tissue. The topi *(Damaliscus korrigum)* is a larger (100–120 kg) antelope that subsists entirely on grasses. Topi select green grass leaves and avoid more fibrous stems and sheaths, thereby giving them 1.5 to 3 times as much protein as they could get eating stem and sheath. Finally, the African buffalo *(Syncerus caffer)* is a large (400–700 kg) grazer that eats a much greater bulk of food than smaller antelope, but forages for it unselectively. Buffalo position themselves in a habitat with as much green grass as possible, and eat most of what they encounter.

For impala, topi, and African buffalo, the microbial farm in the rumen detoxifies most plant allelochemicals. Their feeding ecology can largely be explained by behavioral adaptations that maximize protein consumption and minimize consumption of tough lignified or silicified fiber. What distinguishes the digestive physiologies of these three animals is body size, and with it size of

A

B

C

74

the rumen. Small impala cannot ferment tough foliage easily, so they search out unapparent buds, new leaves, and fruits of apparent shrubs and long-lived grasses. Topi, twice as large as impala, feed on less digestible grasses, but eat only the most digestible parts. Buffalo, a dozen times larger than impala, eat apparent and unapparent tissues almost indiscriminately. However, even massive buffalo cannot digest all fiber (see Table 3-4), and so the fiber and silica contents of their food plants place an ultimate limit on their ability to use the foliage that they consume.

Leaf-cutting ants of the New World tropics provide another example of herbivory in which individuals, and individual colonies, select among a variety of potential foods. In these ants (genus *Atta*), columns of ants from a colony select a herb, bush, or tree suitable for harvest, and carve off pieces of leaves that they carry back to fungus gardens in their underground nests (Fig. 4-11). Like the microbes in a buffalo rumen, the fungus *(Rhozites gongylophora)* degrades fiber and a wide range of allelochemicals. The fungal gardens feed ant larvae, workers, and the single queen responsible for egg production. These ants can be enormously destructive to crops. Abundant *Atta* ants often preclude growing vegetable and orchard crops in tropical regions.

Do *Atta* ants select leaves that they cut among different tree species, or do they feed on whatever leaves are close to their colonies? Stephen Hubbell and David Wiemer (1983) investigated the chemical basis of choice and rejection by matching field observations of foraging ants with chemical assays of potential food plants. Leaves from a forest in Costa Rica were frozen and flown to the United States, where samples of each tree species were ground up and separated into chemical fractions. Oatmeal flakes were dipped into each of the first two fractions, and then were presented to ants from captive colonies. The ants took flakes from one fraction and rejected the other. The rejected fraction went back to the chemistry laboratory for another fractionation, and for repetition of the bioassay. In this way, the chemical basis for rejection could be traced to a single purified chemical compound that makes a tree species unacceptable to ants.

The leaf-cutter bioassay has revealed a great deal about the chemical basis of selection by at least one important generalist herbivore, *Atta cephalotes* (Hubbell et al., 1984). One important finding of Hubbell and his co-workers is that very tough leaves discourage ants. Lignified hemicellulose prevents cutting, while the tannin content of foliage has no influence on ant foraging. The generalist herbivore, like others that eat young tree leaves in Panama (Coley,

Fig. 4-10 Three ruminant herbivores on the Serengeti Plains of Tanzania, East Africa. (A) Impala (40–55 kg) eat buds, new leaves, and fruits of grasses and broad-leaved plants. A small fermentation system and high metabolic requirements force diet selectivity. (B) Medium-sized topi (100–120 kg) eat green grasses and avoid fibrous stems and sheaths. (C) Large African buffalo (400–700 kg) eat and digest both green and fibrous plant parts. A large volume of microbial symbionts and lower metabolic needs per unit weight permit an unselective diet. Photograph of impala by L. K. Johnson of the Smithsonian Institution. Topi and buffalo supplied by Carolina Biological Supply; photographed by William R. West.

Fig. 4-11 Leaf-cutter ants *(Atta)*. Workers cut bits of leaves (top) to feed underground fungal gardens (bottom). Workers supply and tend a fungal garden that the colony, including the egg-laying queen (bottom center), eat. Enormous fungal gardens digest a wide variety of heavily defended plant tissues, but cannot break down fungicidal allelochemicals. Top photograph by Chip Clark, Smithsonian Institution; bottom photograph by Neal A. Weber, Florida State University.

1983), circumvents the digestibility-reducing effects of tannins. Furthermore, choice of tree species is not related to nitrogen content of the leaves, suggesting that protein availability by itself does not govern palatability. This raises the possibility that allelochemicals, rather than gross nutritional content, influence food selection by ants. Toxins might adversely affect the ants, the fungi that the ants cultivate, or both. The toxin hypothesis is clearly confirmed for one legume tree, *Hymenea courbaril,* which is avoided by leaf-cutters. *Hymenea* contains the sesquiterpenoid caryophyllene epoxide, a commercial fungicide, which is lethal to the *Rhozites* fungi that leaf-cutters cultivate (Howard et al., 1988). Many apparent trees seem to have quantitative defenses as predicted by apparency theory, but they supplement them with a remarkable arsenal of qualitative allelochemicals.

Generalist herbivores like savanna mammals and leaf-cutter ants illustrate some of the strengths and limitations of apparency theory. As expected, leaf-cutter ants and all but the largest ungulates are discouraged by silicified or lignified tissues of apparent grasses and trees, respectively. Apparent long-lived plants succeed in using indigestible fiber and silica to limit, to at least some degree, generalized herbivores. Both ungulates and leaf-cutters use a microbial farm, either a gut flora or underground fungal gardens, to detoxify allelochemicals. But antelope and leaf-cutter ants must consequently pick and choose among many plant species, thereby avoiding overdoses of particular plant defensive compounds.

A surprise is that tropical trees, which are as apparent as any English oak, contain many simple toxins in addition to digestibility-reducing compounds like lignins and tannins. Another surprise is that phenols play no part in defense against leaf-cutting ants. Fungal gardens detoxify tannins that would deter or kill English winter moths and many other insects. The presence of toxins in long-lived trees and shrubs indicates that plants pursue mixed strategies of defense, not altogether consistent with apparency theory.

Life on a Host

Many butterfly and moth caterpillars, beetle grubs, fly and wasp larvae, and entire generations of aphids and scale insects never leave an individual host plant. They feed where their mother laid eggs. Sedentary habits that limit challenges to the digestive tract are necessary for most animals with gut cavities too small for extensive fermentation by gut microbes. Both the ecological intimacy of living on a host and small body size promote strong selection for mothers capable of finding the best host available, and for larvae (or adult aphids or scale insects) that can cope with the particular allelochemicals of that host. Apparency theory should apply directly to insects that spend a major part of their lives on one individual or species of host plant.

A survey of published lists of host use tests the prediction the caterpillars are more likely to specialize on unapparent than apparent plants (Futuyma, 1976). Butterflies of the United States support this prediction, but moths are more

Table 4-5 Numbers of butterfly and moth species feeding on more than one family of vascular plants

	Hosts	
	Woody	Herbaceous
Species of Butterflies (eastern United States)		
On 1 family (number)	36	97
On >1 family (number)	17	13
Percentage on 1 family (%)	68	88
Species of Moths (British Isles)		
On 1 family (number)	119	96
On >1 family (number)	100	47
Percentage on 1 family (%)	54	69

Source: Futuyma (1976); original data from various sources.

likely to eat woody plants than herbs (Table 4-5). Within each category (i.e., butterflies or moths), more species feed on only one family of unapparent herbaceous plants, and fewer on apparent woody plants, than would be expected by a random distribution of preferences. These results are generally consistent with apparency theory. But food preferences do not provide a clear-cut vindication of apparency theory because a number of species do specialize on apparent plants.

Feeding-specialization Hypothesis

The idea that plants evolve qualitative defenses that exclude all but specialists has a corollary in animal ecology. The **feeding-specialization hypothesis** predicts that polyphagous herbivores handle the sophisticated chemical defenses of any single food plant less efficiently than monophagous or oligophagous species. Monophagous species, the reasoning goes, should be able to easily degrade toxins in foods that they normally eat.

Tests of the feeding-specialization hypothesis use measures of insect growth, food-processing efficiency, and survival to determine whether some host plants are more suitable for a given insect than others (Appendix II). Tests are most direct when feeding efficiency can be examined with closely related insects on the same food plants. J.Mark Scriber (1983, 1984) provides such a comparison with species of the swallowtail butterfly genus *Papilio.*

Scriber tested the basic hypothesis that specialists use their host plants more efficiently than generalists that eat the same foliage. Spicebush swallowtail caterpillars *(Papilio trolius)* eat only spicebush *(Lindera benzoin)* and sassafras *(Sassafras albidum),* while tiger swallowtail caterpillars *(Papilio glaucus)* have at least 20 acceptable food species. Specialist and nonspecialist caterpillars fare differently on diets restricted to spicebush leaves. Spicebush specialists eat three times the quantity (163 versus 54mg/day/g), have nearly twice the overall effi-

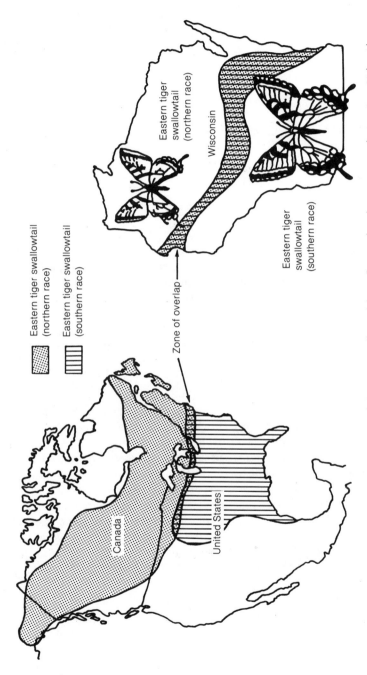

Fig. 4-12 Left: Geographic ranges of subspecies of the tiger swallowtail butterfly (*Papilio glaucus*) in North America. Right: Abrupt boundary between Wisconsin distributions of the northern race (*Papilio glaucus canadensis*) and southern race (*Papilio glaucus glaucus*) of the tiger swallowtail. The northern race cannot survive on some hosts used by the southern race, and vice versa. After Scriber (1983).

Table 4-6 General predictions from apparency theory

	Unapparent Plants	Apparent Plants
Herbivores	Few species	Many species
	Simple interaction	Diffuse herbivory
Selection		
On plants	Strong	Strong
On herbivores	Strong	Strong or weak
Coevolution	Few interacting genomes	Many interacting genomes
	Step-wise	Step-wise or diffuse
Defenses		
Type	Toxins	Toxins and structural digestibility reducers
Against	Generalists	All species
Counteradaptation	Often by specialists	Sometimes by specialists; often by generalists

Source: After Fox (1981).

ciency (21 versus 12%), and have a relative growth rate five times as great (33 versus 6 mg/day/g) as tiger swallowtails from the same locality. In this example the specialist is clearly more efficient on its normal host than is a closely related generalist.

Hypotheses about feeding specialization can also be tested with widespread insect species that eat different plants in different parts of their geographical ranges. Apparency theory would predict that generalists from different localities should be able to eat leaves of the same tree species. Because quantitative defenses are supposed to be general digestibility reducers, local differences in levels of digestibility-reducing defenses should not be extreme enough to select for local specialization. Local specialization would be more likely if apparent species contained toxins with qualitative effects.

The tiger swallowtail is a good species with which to investigate feeding specialization because it is polyphagous and ranges over much of North America (Fig. 4-12). The northern subspecies *(Papilio glaucus canadensis)* eats a variety of northern trees, including quaking aspen *(Populus tremuloides)*. The southern subspecies *(Papilio glaucus glaucus)* eats leaves of trees of a more southern distribution, such as those of the tulip poplar *(Liriodendron tulipifera)*. The geographic boundary between these two butterfly races closely coincides with the climatic limits of several of the food plants of the two populations. Significantly, northern swallowtails die when fed tulip poplar leaves, and southern swallowtails die when fed quaking aspen. Inherited local differences in swallowtail feeding efficiencies also occur on other food plants.

Regional and local host specialization may be common among insects (Fox and Morrow, 1981). Widespread insect species sometimes appear locally specialized because only one of several food plants is common in any given locality. But genetic studies of feeding efficiencies show that real feeding specialization does occur among polyphagous insects. Regional specialization on apparent woody plants is not predicted by apparency theory. This fact, in addition to the repeated discovery that apparent plants use both toxins and digestibility reducers, indicate the need for a revision of apparency theory (Table 4-6). As often happens in science, theory must be adjusted to fit the facts.

RESOURCE AVAILABILITY THEORY

Defensive capabilities of plants are constrained by the resources available to them. Phyllis Coley, John Bryant, and F. Stuart Chapin III (1985) propose that plant defensive capabilities are mediated by their capacity to replace lost parts with resources at their disposal, rather than by apparency to herbivores. While apparency theory implies that herbivore foraging efficiency strongly influences plant defense, resource availability theory assumes that inherent growth physiology, photosynthetic capability, and nutrient availability determine the amounts and kinds of defenses that plants use.

Resource availability theory assumes that the conditions under which inherent growth potential is expressed determine plant defense. The three variables that determine the amount and kind of defense investment are (1) inherent growth rate, (2) overall resource availability, and (3) carbon/nitrogen balance. These three variables interact to determine whether investment in defenses is high or low; whether the defenses employed are immobile fixtures of plant tissues (e.g., lignins, tannins), or are mobile metabolites that are constantly being created and destroyed (e.g., terpenoids, alkaloids); and whether the defenses employed are carbon-based (e.g., lignins, tannins, terpenoids) or nitrogen-based (e.g., alkaloids, cyanogens, nonprotein amino acids).

Growth Rate and Defense Investment

The first issue is the relationship between inherent growth rate and overall investment in defenses (Fig. 4-13). Plant species that grow rapidly in well-lighted environments with fertile soil can easily replace leaves or other tissues lost to herbivores. Such plants are usually early successional trees and herbs. Coley and her colleagues predict that these rapidly growing species will invest relatively little in defenses of any kind, and the defenses employed will be easily metabolized mobile molecules. Rapidly growing species or tissues minimize investment in lignins or condensed tannins that might inhibit tissue elaboration. At the other extreme, slow-growing plants invest more in defenses with complex polymers such as lignins and tannins. Slow growth favors the use of these immobile defenses which, once employed, are a permanent fixture of leaves and stems.

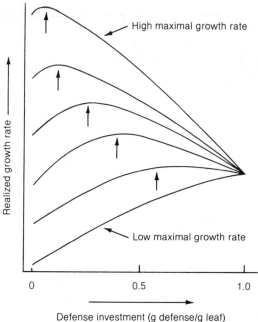

High maximal growth rate

Realized growth rate

Low maximal growth rate

0   0.5   1.0

Defense investment (g defense/g leaf)

Fig. 4-13 Relationship of investment in plant growth and defense according to resource availability theory. Each curve is a plant species with a different inherent growth rate. Slow-growing plants allocate a proportionately greater share of resources to defenses, which can be carbon-based, nitrogen-based, or both depending on the availability of resources for the plant. Rapidly growing plants can easily replace tissues lost to herbivory, and consequently allocate proportionately less to defense. After Coley et al. (1985). Copyright © 1985 American Association for the Advancement of Science.

Cost of Mobile and Immobile Defenses

The resource availability theory differs from plant apparency theory in its assessment of the energetic cost of immobile (quantitative) and mobile (qualitative) defenses. Because individual polymers have more atoms than smaller toxin molecules and are found in high concentrations in plant tissues, plant apparency theory assumes that quantitative defenses that depend on polymers cost more for plants to use than qualitative defenses. This is not necessarily accurate if immobile quantitative defenses are made only once, while the smaller mobile qualitative defenses are constantly being synthesized and broken down (Fox, 1981). If the rate of synthesis and breakdown (turnover rate) of mobile alkaloids or terpenoids is sufficiently high, their use may require more energy overall than those of much larger and more prevalent polymers.

Resource availability theory assumes that mobile toxins cost more to make than digestibility reducers for long-lived tissues. Quantitative defenses are advantageous when overall resource limitation imposes a slow growth rate on a species, and when tissues need not be constantly replaced. Qualitative toxins

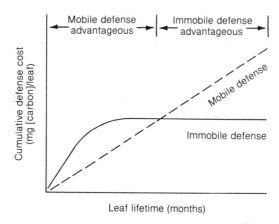

Fig. 4-14 Advantages of defense by mobile toxins as compared with immobile digestibility reducers in the life of a leaf. The cumulative cost in defense increases indefinitely with age for mobile defenses, such as alkaloids or terpenes. This is not true for immobile digestibility reducers. A large initial investment in these complex polymers will suffice for the life of a leaf. Mobile toxins should be most efficient for rapidly growing, short-lived, plant tissues, while immobile digestibility reducers should be most efficient for slow-growing, long-lived leaves and stems. After McKey (1984) and Coley et al. (1985).

are an advantage for fast-growing species and tissues because they are flexible. A rapidly growing plant or tissue can metabolize mobile defenses, or translocate them within the plant, as needed. If leaves are lost to herbivores, replacement leaves can be elaborated and quickly defended.

Resource availability theory can be used to explain changing patterns of defense in maturing tissues (Fig. 4-14). Young rapidly growing tissues rely on mobile defenses, while older tissues rely on immobile digestibility reducers. This explains why buds and young leaves are often impregnated with alkaloids, terpenes, and other toxins (see Fig. 4-7), or are defended by literally mobile ants (McKey, 1984). For resource availability theory, the economics of growth, rather than apparency to herbivores, determines investment patterns.

Carbon/Nitrogen Balance

Resource availability theory predicts that the kind of nutrients available determine whether defenses are carbon-based or nitrogen-based. Plants growing on nitrogen-poor soils will use carbon-based toxins such as terpenoids as mobile defenses, while those growing on fertile soils are more likely to use nitrogen-based toxins such as alkaloids or cyanogens as mobile defenses. Slow-growing plants living under low light availability on poor soils have long-lived leaves that are most economically defended with immobile polymers such as carbon-based tannins and lignins.

RESOURCE AVAILABILITY OR PLANT APPARENCY?

Can growth physiology and resource availability explain patterns of plant defense better than apparency to herbivores? Can predictions from resource availability theory be borne out by observation (Table 4-7)? Although apparency theory does explain some patterns of herbivore specialization, the new theory accommodates other patterns of plant defense better than apparency theory.

Patterns of mammalian herbivory that are tangential to apparency theory are accommodated by resource availability theory. The roots of plant resource theory lie in studies of snowshoe hare *(Lepus americanus)* herbivory in Alaska. John Bryant and his associates (1980, 1983, 1985) noticed that trees growing in nitrogen-poor soils lacked familiar nitrogen-based defenses, such as alkaloids, but were well-defended with carbon-based toxins. Green alder *(Alnus crispa),* for instance, defended its young shoots with a carbon-based toxic phenol, pinosylvin methyl ether. In another example, Reichardt, Bryant, and others (1984) found that snowshoe hares were repelled by a carbon-based terpene, papyrific acid, in growing shoots of the Alaska paper birch *(Betula resinifera).* Older alder and birch tissues were protected by resins and lignins. A theory emphasizing resource availability has the pleasing property of explaining both the differences in chemistry between young and older tissues, and the reasons that allelochemicals of younger tissues in plants that grow on nitrogen-poor soils are carbon-based toxic phenols and terpenes rather than more familiar alkaloids or cyanogens.

Another strength is that high rates of herbivory on early successional foliage or young plant tissues are predicted in resource availability theory, which assumes that rapidly growing but poorly defended tissues can be easily replaced. Apparency theory has no predictions on the issue, explaining such occurrences as the chance failure of unapparency.

Will resource availability theory accommodate all relevant patterns in nature? It is far too early to tell. For one thing, resource availability theory emphasizes the plant perspective, much as apparency theory emphasizes the animal perspective. Resource availability clarifies the mechanisms that plants must use to meet challenges by plant-eating animals, but it has not yet offered any predictions on such basic issues as host specialization by herbivores, or coevolution of herbivores and plants.

Are the resource availability and apparency theories exclusive? Both theories share some predictions, such as the expected prevalence of toxins in early successional plants, and digestibility reducers in late successional species. A botanist primarily interested in plant physiology might see apparency theory as a subdivision of resource availability theory. To a botanist, the general theory is one of plant resource allocation, while causes of host use or avoidance by particular herbivores are subsidiary. To an entomologist, plant apparency might provide more insight into host specialization and coevolution (Table 4-6), while a subsidiary theory of resource allocation defines the constraints on plant re-

Table 4-7 Growth and defense of plants according to
resource availability theory

Variable	Fast-Growing Species	Slow-Growing Species
Growth Characteristics		
Normal resource availability	High	Low
Maximum growth rate	High	Low
Maximum photosynthetic rate	High	Low
Dark respiration rate	High	Low
Leaf protein content	High	Low
Responses to pulses in resources	Flexible	Inflexible
Leaf lifetimes	Short	Long
Successional status	Often early	Often late
Antiherbivore Characteristics		
Rate of herbivory	High	Low
Ability to replace lost tissue	High	Low
Defense investment	Low	High
Type of defense (sensu Feeny)	Qualitative	Quantitative
Turnover rate of defense	High	Low
Flexibility of defense expression	More	Less

Source: After Coley et al. (1985). Copyright © 1985 American Association for the Advancement of Science.

sponses to herbivory (Table 4-7). With time the glare of the scientific spotlight will illuminate important distinctions between these theories and either show which explains both plant defense and host use best, or show how they can be assimilated into a general theory of plant defense and herbivore use.

SUMMARY

Ecological relationships of plants and the animals that eat them are conditioned by the foraging and digestive efficiencies of the eaters, and the defensive capabilities of the eaten. Apparency theory suggests that long-lived trees, shrubs, and grasses cannot escape being discovered by herbivores. Consequently, they are well-defended by lignified, cutinized, and silicified cell walls or woody tissues that impede digestion, as well as by a wide array of toxins that are effective in low concentrations. Short-lived herbs or woody plants of early successional stages rely heavily on toxins that are poisonous or lethal in very small quantities, except against specialists that have counteradapted to them.

Resource availability theory clarifies the mechanisms by which plant defense is possible. Resource availability in the environment and inherent growth rates are major determinants of the amount and kind of defenses employed by plants. Especially important is the hypothesis, not yet widely confirmed, that high turnover of small toxin molecules imposes a higher energetic cost on plants (and presumably a drain on fitness in the *absence* of herbivores) than complex digestibility-reducing polymers.

Different kinds of herbivores have entirely different ecological relationships with their food plants. Large, polyphagous mammals live in a fine-grained world of many kinds of food plants. Generalized enzyme systems and gut microbes allow them to disarm most toxins. Diets of polyphagous herbivores are largely limited by the proportion of lignified or silicified tissues in food plants. Insects and other small herbivores live in a coarse-grained world in which toxins severely restrict food choice. Differences of scale in size, longevity, and digestive constraints make the same forest very different worlds to a caterpillar and a deer.

Differences in environmental grain translate into different possibilities for coevolution between plants and animals. Animals that sample hundreds or thousands of plants in a lifetime are not likely to evolve particular adaptations to a single plant species. Nor is a plant species that is nibbled by many animals likely to evolve separate defenses against each one. Herbivore specialists do impose strong selective pressures on plants, and plants often respond with novel allelochemicals elaborated from previously used plans, much as Ehrlich and Raven (1964) envisioned over 20 years ago. More complex theories of plant apparency to herbivores and plant capacity for defense are needed to explain widely divergent patterns of herbivore specialization, and of plant defense.

STUDY QUESTIONS

1. An ecologist steeped in predator/prey theory notices that the Earth is green. He argues that plant success shows that herbivores are held in check by predators and disease, otherwise animals would consume all of the vegetation on the planet. What is wrong with this argument?

2. A biologist concludes that tannins are not toxic because the monkey that he studies eats tannin-rich leaves. Should the biologist eat the leaves?

3. Suppose a leaf-cutter ant bioassay fails to find a single compound that deters ants from some tropical tree. Instead, the last two fractions of the chemical extraction are palatable to ants, but the previous fraction, which produced them, is repellent. How would you explain such results?

4. A one-inch-long marine snail eats algae from rocks along the seashore. Is its pattern of herbivory most similar to that of an aphid, a grasshopper, or a horse?

5. Under what circumstance might a qualitative toxin, such as an alkaloid or terpenoid, act in a dosage-dependent manner?

6. Some biologists believe that somatic mutations in long-lived plants are adaptive because they result in variable chemical defenses among different branches of a single plant. How could such adaptive mutation evolve by natural selection?

7. Compare and contrast plant apparency and resource availability theories of plant defense.
8. What would have to be incorporated into a general theory of plant defense and herbivore specialization?
9. John Thompson (1986) notes that taxonomic uncertainties often make it difficult to interpret host use by insects. Some taxonomists favor placing many species in a genus, while others elect to create many genera with few species in each. How might such taxonomic biases influence library tests of theories of plant defense?

SUGGESTED READING

Insect herbivory is well-covered by Michael Crawley (1983) in *Herbivory,* by Robert Denno and Mark McClure (1983) in their edited book *Variable Plants and Herbivores in Natural and Managed Systems,* and by Donald Strong, J. H. Lawton, and Richard Southwood (1984) in *Insects on Plants.* A. R. E. Sinclair and M. Norton-Griffiths (1979) take a broad view of mammalian ecology and herbivory in *Serengeti: Dynamics of an Ecosystem.* Space limits discussion here to herbivory in terrestial communities. Jane Lubchenco and Steven Gaines (1981, 1982) provide access to the literature on aquatic and marine herbivory in two overviews in *Annual Review of Ecology and Systematics.*

5

Evolution and Herbivory

What are the evolutionary consequences of ecological interactions between animals and plants? **Coevolution** is the simultaneous evolution of ecologically interacting populations (Roughgarden, 1983). For example, a plant might evolve a novel secondary compound that only a few individuals of its insect enemies can detoxify. If those fortunate insects pass resistance on to their offspring, genes for resistance will spread in the herbivore population and the early advantage to the new toxin erodes. Herbivores gain the upper hand until a variant allelochemical appears to renew the coevolutionary cycle. In principle, such a reciprocating chemical arms race can proceed indefinitely.

Pairwise coevolution in which two species adapt specifically to each other is not the only possible or likely outcome of ecological interaction. Diffuse coevolution occurs when two sets of species interact, each set influencing the other more or less equally. Alternatively, asymmetrical evolution occurs without coevolution. An insect evolves and its host does not, or vice versa. Asymmetrical coevolution occurs if the host or insect evolves more rapidly than its opposite. Often, interaction without evolution leaves herbivores and their food plants genetically unchanged. What determines which of these alternatives occurs?

Modes of inheritance determine whether and how plants and herbivores respond to the ecological challenges that they impose on each other. Plant defense or resistance to it may be conferred by one or many genes, and those genes may influence one or many traits in the plant or animal. This chapter explores the genetic bases of plant defense and herbivore resistance.

GENETIC EVOLUTION

Darwin and other early evolutionists thought that natural selection caused continuous and inexorable adaptive change. It is now clear that evolution may be continuous or discontinuous, and that the direction of evolution frequently shifts with changes in ecological conditions. ➤

Natural selection has different consequences, depending on the part of a population being selected (Fig. 5-1). Directional selection favors one end of the phenotype distribution, resulting in a shift of allele frequency (or frequencies)

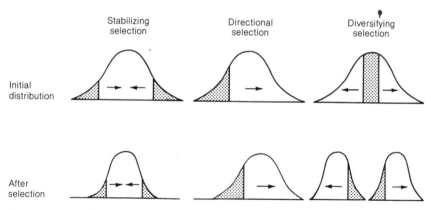

Fig. 5-1 Modes of natural selection. The curves represent the distributions of phenotypes of populations. Shading represents the phenotypes *culled* by natural selection. Top: Initial distributions at the onset of directional selection, stabilizing selection, and disruptive or diversifying selection. Bottom: Resulting distributions as selection proceeds. Arrows indicate the direction of phenotypic change under selection.

in favor of genotypes that produce the favored phenotype(s). The direction of selection may change from one time or place to another. Stabilizing selection favors intermediate phenotypes at the expense of extreme phenotypes. It preserves the status quo of average phenotypes, but reduces variance. Disruptive selection favors extreme phenotypes at the expense of intermediates, sometimes leading to local genetic adaptation. These different modes of selection occur in both single-gene and polygenic traits.

The *time scale* of genetic evolution required to fix or eliminate genes may be vast, as compared to the time scale of some ecological changes. Suppose, for example, that a dominant allele *A* has a 10% advantage in fitness over the recessive *a* (Appendix I). It would take over 100 generations for a new mutant *A* to reach a frequency of 90%. (Fig. 5-2). The recessive *a* would not be eliminated in 1,000 generations. If instead the recessive gene had a 10% advantage, it would take nearly 1,000 generations for the frequency of a new mutant *a* to reach 90%. One thousand generations for an annual plant is 1,000 years. For a tree, 1,000 generations could require 20,000 to 50,000 years, or most of a geological ice age.

The time scale required for substantial genetic evolution is important in an ecologically dynamic world. Idealized selection with constant selection intensity (Fig. 5-2) rarely, if ever, occurs in animal and plant relationships for long. Idealized simulations are most useful for predicting intense selection for short periods, as might occur when a plant population is faced with a new pest, or when insect pests respond to a new insecticide. In nature the intensity of selection is often lower, or frequently shifts in direction. Local succession and regional climatic change alter the distribution and abundance of competitors, food plants, and herbivores, thereby changing selection coefficients for single gene traits and selection intensities for polygenic traits (Appendix I). Natural selec-

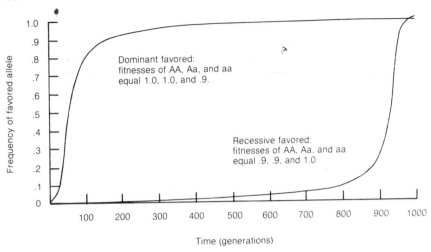

Fig. 5-2   Change in the frequency of a favored dominant (top curve) and a favored recessive (bottom curve) allele. Note that the frequency of a favored dominant changes rapidly when it is rare and slowly when it is common. The frequency of a favored recessive changes most slowly when an allele is rare, and much more rapidly when it is common and therefore not usually masked by the dominant allele. After Hartl (1981).

tion on an allele or polygenic set of alleles may be directional in one decade, century, or millenium, and stabilizing or disruptive in others.

The simple inexorable directional change implied by Darwin in *The Origin of Species* is not the only, or even the most common, consequence of natural selection. Simple Mendelian traits with strong effects in herbivores and plants might coevolve in a stepwise reciprocating fashion, as predicted by the classical theory of biochemical evolution. But other modes of selection and response to it might be at least as important. Polygenic traits in plants could evolve at quite different rates from those in animals, leading to strongly asymmetrical adaptation of plant resistance, or pest virulence. The dynamism of natural selection in a variable world implies a variety of possible patterns of herbivore and plant adaptation and counteradaptation.

COEVOLUTION IN AGRICULTURE

The most complete genetic models of herbivore and plant coevolution are in agricultural systems. Agronomists create strains of crop plants that are resistant to insect and microbial pests. The genetic basis of host resistance is known for many crop species afflicted with particular insect pests, and some information is available concerning the genetics of host use by insects (Table 5-1). These agricultural examples suggest that a wide variety of genetic mechanisms are likely to underlie relationships between insects and herbivores in nature.

Table 5-1 Selected examples of genetic variation influencing resistance to herbivores by plants, and resistance to plants by herbivores

Plant	Herbivore	Plant Genetics	Herbivore Genetics	Plant Resistance
Alfalfa	Alfalfa aphid	Polygenic	6 biotypes[a]	Antibiosis, nonpreference, tolerance
Cucumber	Spider mite	1 dominant allele and modifiers	Polygenic	Antibiosis, nonpreference
Raspberry	European raspberry aphid	9 dominant alleles at different loci with independent effects	Several single dominant alleles at different loci	Antibiosis, nonpreference
Rice	Brown planthopper	2 dominant, 2 recessive alleles at different loci with independent effects	Polygenic	Antibiosis, nonpreference
	Green leafhopper	4 dominant, 1 recessive alleles at different loci with independent effects	2 biotypes	Antibiosis
Sorghum	Corn-leaf aphid	1 dominant allele	5 biotypes	Antibiosis
Wheat	Hessian fly	7 dominant, 1 recessive alleles at different loci with independent effects	4 recessive alleles at different loci with independent effects	Antibiosis

Source: After Gould (1983), derived from many original sources.

[a]The genetic basis of a biotype may be a single gene, linked genes, or polygenic.

   The genetic basis of resistance to insects can involve one locus, several independent loci that confer resistance in different ways, or several or many loci that are responsible for a polygenic phenotype (Table 5-1). Sorghum protects itself against the corn-leaf aphid with one dominant allele, while alfalfa resistance against the alfalfa aphid is polygenic. Furthermore, the same plant species may be protected against different herbivores in rather different ways. The resistance of rice to brown plant hoppers depends on either of two dominant or two recessive alleles at four loci, while its resistance to green leafhoppers is conferred by either of four dominant or one recessive allele(s) at different loci. Raspberry defense against European raspberry aphids is complex, but defense against the American raspberry aphid depends on only one dominant allele. The genetic basis of host resistance is clearly varied.

Plant resistance to insect pests takes one of three forms, each of which may involve single gene or polygenic inheritance. **Antibiosis** kills or diminishes the fitness of insect pests, either by mechanical or biochemical means. If an insect eats a plant, it is sickened, killed, or suffers a loss in reproductive capacity. **Nonpreference** either changes chemical properties of the plant to an herbivore, so that it is not perceived, or discourages oviposition. Plant resistance that results in nonpreference keeps an insect enemy from occupying the host. **Tolerance** means that the plant experiences infestation without losing fitness or, in agricultural systems, "yield." Some plants have more comprehensive resistance than others (Table 5-1). Rice resistance to leafhoppers is antibiotic, while resistance to planthoppers involves both antibiosis and nonpreference. Polygenic resistance of alfalfa to the alfalfa aphid includes antibiosis, nonpreference, and tolerance.

Finally, herbivore resistance to host defenses is varied. **Biotypes** are strains of insects with inherited differences in their ability to use host species. Biotypes have traditionally been identified by insect use of particular host varieties, without a genetic analysis of the insect characteristics. Recent genetic analysis shows that some biotypes are single-locus effects, while others are caused by closely linked genes that are passed on without recombination. Others turn out to be polygenic. The term biotype is worth preserving for the time being because most strains of crop pests are identified as biotypes without genetic analysis. As the different genetic mechanisms conferring resistance to plant defenses become better known, the term will probably fall into disuse. Perhaps the most important general lesson is that the genetic basis of an insect biotype need not match the genetic basis of plant resistance. For instance, spider mite resistance against a single dominant cucumber gene is polygenic.

No random sample of systems exists that can indicate which mechanisms are most common, or how plant adaptation and insect counterdaptation usually occur in nature. Agricultural examples do suggest what is *likely* in nature, and point to important directions for future investigation.

Coevolution and Mendelian Traits

Agronomists often search for traits with clear Mendelian effects on host resistance, and clear Mendelian responses from pathogens or insects. H. H. Flor (1956) described just such an effect with flax *(Linum usitatissimum)* and a rust fungus *(Melampsora lini)*. Resistance in flax is governed by 27 alleles distributed among five loci. Virulence in the rust is conferred by a complementary system in which each resistance gene in the flax is countered by a virulence gene in the fungus. One-to-one correspondence of Mendelian host and pest genes is called **gene-for-gene coevolution,** and has been widely applied to host–fungus coevolution. The gene-for-gene concept predicts that each plant gene that confers defensive capability is countered by a distinct corresponding insect gene that confers a clear-cut ability to use a given host strain.

The best example of the gene-for-gene concept in insect and plant interactions is the case of the Hessian fly *(Mayetiola destructor)* in wheat *(Triticum aestivum)* (Gallun and Khush, 1980). Hessian flies overwinter as larvae in pupal cases under the leaves of winter wheat. When the adults emerge, they lay their eggs on the wheat, and the larvae feed between the leaf sheath and the stem. Susceptible wheat stems are weakened, stunted, and sometimes killed.

Wheat has seven dominant genes and one recessive gene that confer resistance to the Hessian fly. Four of these dominant genes *(H$_1$, H$_2$, H$_3$, and H$_4$)* are matched, one on one, by recessive genes that confer resistance in the fly *(aa, bb, cc, dd)* to the plant. The homozygosity of recessives required for resistance is aided by a peculiar meiotic system in which male Hessian flies pass only the *maternal* chromosomes to offspring. Hence males always breed as if they were homozygotes. The consequence for agriculture is that Hessian flies have rapidly evolved resistance to several wheat strains without losing the ability to evolve resistance to additional wheat defenses.

Is gene-for-gene coevolution common? All biotypes were once thought to represent gene-for-gene systems, but this assumption is now controversial. For instance, the brown planthoppper *(Nilaparvata lugens),* which dehydrates and kills rice, exists in several biotypes that for many years were considered simple Mendelian traits. M. F. Claridge and his co-workers (1980, 1982a,b) have shown that planthopper biotypes are actually locally adapted polygenic phenotypes. The planthopper biotypes freely interbreed, overlap broadly in their resistance properties, and can be created from one another in a few generations by natural or artificial selection. These polygenic phenotypes quickly evolve (i.e., become new biotypes) when exposed to a new resistant rice variety.

Selection on Polygenic Traits

There is a growing consensus that host use by insects is often polygenic, rather than attributable to simple Mendelian inheritance. Conceptually, this means that biologists must distinguish genetic and environmental components of phenotypic variation in host use (Appendix I). In practice, polygenic inheritance implies that local biotypes, or even progeny lines of insect pests, vary in preference for, or virulence to, host plants.

Recent explorations of polygenic inheritance of host use suggest active and continuous adaptation among phytophagous insects. *Lyriomyza sativae* is a fly that infests many vegetable crops. Adult flies lay their eggs in leaves, and the maggots feed by burrowing through the leaf tissues. Sara Via (1984) studied the genetics of host use in this leaf miner by raising members of the same families (offspring from the same parents) on tomato *(Lycopersicon esculentum)* and cowpea *(Vigna unguiculata)* plants. She found wide variation in development time among families raised on different hosts (Fig. 5-3). Some fam-

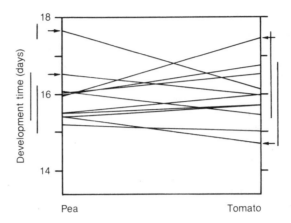

Fig. 5-3 Variation in development time among families of leaf-mining flies *(Lyriomyza sativae)* on peas and tomatoes. Lines connect the means of siblings raised on each host. Differences in elevations of means indicate genetic variation within families in the ability to use the two hosts. Nonparallel lines indicate genotype/environmental interaction in development time. After Via (1984).

ilies developed more rapidly on tomatoes, others developed more rapidly on peas. The rank of family development time on one host did not correspond to its rank on the other. Via showed that individual flies were potentially generalists in food choice, but that polygenic variation in resistance predisposed some to do better on certain hosts than others. Similar genetic variance in host preference in fruit flies *(Drosophila tripunctata)* suggests that a polygenic basis for host choice may be widespread in nature (Jaenike, 1985).

Crop pests evolve monogenic or polygenic resistance to novel plant strains within 8–12 years of being exposed to them (Maxwell and Jennings, 1980). Evolution of resistance, for either plants or insects, may be slower in nature. Crop plants are usually unapparent species made apparent by monocultures that maintain them in unnaturally high densities over vast areas for decades or even centuries. Compared with herbivorous insects in fields and forests, crop pests find it easier to locate hosts, are less likely to die searching, and exist in artificially immense populations. Plants with a novel toxin are embedded in a huge insect population that is likely to contain mutants capable of overcoming the defense. We expect, and see, rapid evolution of resistance to crop plants. What happens in nature?

EVOLUTION AND HERBIVORY IN NATURE

The genetic basis of coevolution of plants and herbivores in nature is not well understood. Evolutionary ecologists must infer the likely course of coevolution, given possible genetic contexts and known ecological circumstances.

Predictable and Variable Food Plants

Evolution of herbivores and plants is influenced by the consistency with which both the herbivore and the plant interact with each other. For small herbivores living on a host plant, the environment is so coarse-grained that specific adaptations to a host species, or even to a host individual, are possible. For a polyphagous herbivore that eats many plants, the environment is so fine-grained that diet selection and generalized detoxification systems are likely to play a larger role than biochemical adaptation to particular food plant species.

The black pineleaf scale *(Nuculapsis californica)* is a minute sapsucking insect that infests pines (genus *Pinus*). *Nuculapsis* females are wingless and legless. Each female spends its sedentary adult life attached to a pine needle. Winged males lack mouthparts and do not eat. Mobility is limited, occurring during the first larval phase in which minute "crawlers" colonize neighboring trees or different parts of the tree on which they were born. With a generation time of 1 year, dozens of *Nuculapsis* generations may occur on a single ponderosa pine *(Pinus ponderosa)*. Scales can be serious pests of pines, reaching densities of 10 per centimeter of needle.

The intimacy of the relationship between the short-lived scale and its long-lived host is a classic example of coarse-grained herbivory. George Edmunds and Donald Alstad (1978) provide convincing evidence that scale colonies evolve the means to overcome defenses of the *individual* pines on which they live. Some pines have much heavier infestations than others. Twigs taken from one tree and grafted to another retain their ability (or inability) to prevent colonization by pineleaf scales. Edmunds and Alstad hypothesize that over many generations colonies of scales adapt to the unique individual combinations of terpenoids, resins, quinones, and other allelochemicals of their long-lived hosts. Strong support for this hypothesis comes from the high survival rate of scales transferred to branches within the same tree, as compared with those that are transferred to neighboring trees (Fig. 5-4). Trees, in effect, become progressively vulnerable as successive generations of scales overcome one allelochemical defense after another.

The black pineleaf scale shares an end of an adaptive continuum with other short-lived herbivores, such as nematodes, aphids, and mealybugs, which live for several to many generations on one host individual or clone. Theirs are coarse-grained worlds in which lifetime feeding challenges are determined at or shortly after birth. At the other end of the continuum is the fine-grained existence of large grazers that eat several plants per minute. One well-studied system lies in between.

English meadows contain numerous grasses and forbs that sustain a diverse community of pathogens, insects and other invertebrates, and plant-eating mammals. An important component of these communities is the clover *Trifolium repens,* which is noted for a polymorphism for cyanogenic activity (Table 5-2). Cyanide-producing clovers have a dominant allele for the cyanogenic glucoside (allele *Ac)* as well as an unlinked dominant allele for a β-glucosidase enzyme (allele *Li*) capable of releasing cyanide (HCN) gas. Cyanogenic and

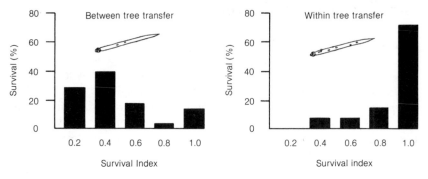

Fig. 5-4 Survival of pineleaf scales *(Nuculapsis californica)* transferred within and between individuals of ponderosa pine *(Pinus ponderosa)*. Far fewer survive transfer between trees than transfer within the same tree, showing that scales are physiologically and perhaps genetically incapable of living on trees other than the one on which they are accustomed to feed. The survival index on the abscissa eliminates the effects of unequal establishment of scale transfers when the experiment was started. After Edmunds and Alstad (1978). Copyright © 1978 American Association for the Advancement of Science.

acyanogenic clovers often occupy different habitats (Jones et al., 1978), but sometimes occur in the same place. Where the clover morphs co-occur, herbivores have a choice of eating toxic or nontoxic forms.

Herbivore preferences are not as clear-cut as one might expect. Rudolfo Dirzo and John Harper (1982a, b) demonstrated that cyanide toxicity of clovers affects four mollusks: three slugs and one snail. Slugs and snails that are fed acyanogenic clovers hold their own in weight, but those fed cyanide-producing forms lose weight rapidly (Fig. 5-5). Slugs and snails in the laboratory much prefer acyanogenic leaves to leaves of the cyanogenic morph, and a field survey shows disproportionate damage to acyanogenic leaves (Table 5-3). But mollusks also eat many cyanogenic leaves. Protection through cyanogenesis is partial, not complete.

Table 5-2 Phenotypes and genotypes of the cyanogenesis polymorphism in the clover *Trifolium repens*

Alleles	Plant Contents	Notation	Reaction
Ac ? Li ?	Cyanogenic glucosides and enzyme	AcLi	Cyanogenic
Ac ? li li	Cyanogenic glucosides but no enzyme	Acli	Acyanogenic
ac ac Li li	Enzyme but no cyanogenic glucosides	acLi	Acyanogenic
ac ac li li	No enzyme and no cyanogenic glucosides	acli	Acyanogenic

Source: Dirzo and Harper (1982a).

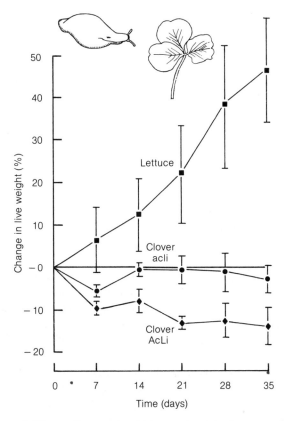

Fig. 5-5 Effects of different diets on slug *(Arion ater)* weight. Leaves include lettuce and cyanogenic (AcLi) and acyanogenic (acli) morphs of the clover, *Trifolium repens*. After Dirzo and Harper (1982a).

Table 5-3 Mollusk herbivory on cyanogenic (AcLi) and acyanogenic (acli) morphs of the clover *Trifolium repens*

Genotype *(N)*	Category of Mollusk Damage (percentage of leaf area eaten)			
	0	<5	5–50	50–100
Field survey of damage by all snails and slugs (% in each category)				
AcLi (671)	40	12	18	30
acli (610)	37	9	13	41
Laboratory test of the slug *Arion ater* (% in each category)				
AcLi (430)	62	3	6	29
acli (303)	9	2	3	86

Source: Data from Dirzo and Harper (1982a).

Table 5-4 Censuses of mollusk density and distributions of
cyanogenic and acyanogenic morphs of the clover
Trifolium repens

	Plant Phenotype	
Mollusk Density	Cyanogenic	Acyanogenic
High to very high	10	1
Low	29	16
Very low	10	27

Source: After Dirzo and Harper (1982a).

Why do slugs sometimes eat cyanogenic clovers? Several possibilities exist. The toxicity of clovers with the AcLi genotype might vary if growing conditions affect their ability to manufacture cyanogenic glucosides. Individual slugs might also vary in their capacity to detoxify HCN. Alternatively, mollusks might simply taste toxic clovers by accident if they happen to be common in habitats where the toxic clover morphs are common (Table 5-4). In fact, mollusks prefer moist situations where cyanogenic clovers co-occur with other food plants. Slugs and snails may taste cyanogenic forms because they come into contact with cyanogenic morphs in moist meadows much more often than with the acyanogenic clovers that they prefer.

Have scale insects and pines, and mollusks and clovers, *co*evolved? Pines vary in allelochemicals that influence many herbivores, while scales adapt to some pines and not to others. Edmunds and Alstad conclude that scales evolve resistance to some pine hosts, but that pines do not adapt specifically to scales. The "co" in coevolution is elusive because this relationship is asymmetrical. Cyanogenic capacity in clover may now be evolving in response to herbivory, but Dirzo and Harper doubt that it evolved in response to long-term herbivory by slugs. If it had, natural selection would have favored close genetic linkage between *Ac* and *Li* loci so that both would invariably be inherited together. The fact that these loci are *not* linked implies that (1) cyanogenesis is a recent trait in clover, (2) the plant has only recently needed β-glucosidase to protect itself (mollusks are a recent challenge), or (3) the capacity to produce both cyanogenic glucosides and β-glucosidase evolved for other reasons. Dirzo and Harper note that a number of other pathogens and herbivores use clovers, and that some actually prefer cyanogenic forms.

Origin of Host Associations

Ecological theory suggests that coevolution is possible, but that it is often unlikely. Differences in generation times between insects and most plants preclude very tight coevolutionary reciprocity. Nor would one expect symmetrical *co*evolution in fine-grained environments in which plants were eaten by highly polyphagous animals. A coevolutionary relationship is most likely between plants and insects that live on or in their tissues.

Genetic theory suggests that many kinds of host adaptation are possible if the important traits are Mendelian (Levene, 1953). Fine-grained environments maintain genetic variation among herbivores indefinitely if heterozygotes are more fit than homozygotes in different host plant environments. The same is true of coarse-grained environments if insect densities are regulated independently on different hosts. If homozygotes are favored over heterozygotes, or insects on some hosts contribute more offspring (alleles) than those of other insect parents to the total species gene pool, the favored allele gains over others and host specialization evolves. Via and Lande (1985) have recently theorized that selection on polygenic traits can similarly lead to either specialization or polyphagy, depending on the sorts of genetic combinations favored by selection and the mobility of the species to be selected. A detailed consideration of these theoretical models is beyond the scope of this book, but the models do indicate that different conditions can lead to, or preclude, specialization of phenotypes influenced by either Mendelian or polygenic genotypes.

Another approach is to ask to what degree the phylogenies of plants and insects are mirror images of each other. If insects and plants coevolved in a reciprocal stepwise fashion, as originally suggested by the classical theory of biochemical coevolution, the phylogenetic relationships of insects should closely reflect the phylogenetic relationships of their plant hosts. This occurs to a re-

Fig. 5-6 Summary of butterfly phylogeny and food plants. Slash indicates close taxonomic affinity; arrow indicates secondary feeding habits. After Mitter and Brooks (1983).

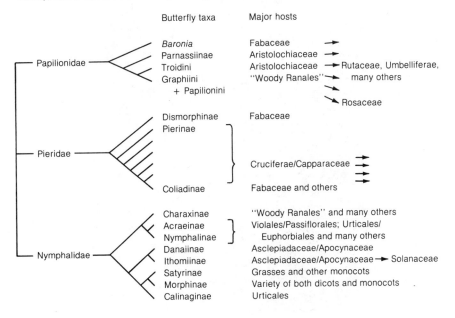

markable degree in the evolution of intestinal parasites; each primate genus has a distinctive species or genus of pinworm (Mitter and Brooks, 1983). Butterfly families and their larval host plants show much less complementarity (Fig. 5-6). Insects diversify on host species that share similar chemistry. In some cases, those hosts are closely related (e.g., subfamily Satyrinae on grasses), as is predicted by the classical theory of biochemical evolution. In others, insects undergo radiations on unrelated plants with convergent chemical defenses. For instance, swallowtail tribes (Papilionini) have adapted to several unrelated plant groups.

The known allelochemistry of some plants suggests that insect radiations often reflect host chemistry more than host taxonomy (Berenbaum, 1983). For instance, the leaf-rolling moth *Agonopterix* avoids light-activated furanocoumarins in parsnips (family Umbelliferae) by rolling and thereby shading the tissues on which it feeds. The same moth genus also rolls, shades, and eats leaves of plants in the sunflower family (Asteraceae), bean family (Fabaceae), and citrus family (Rutaceae), all of which contain members with furanocoumarins (Fig. 4-4). *Agonopterix* even feeds on one largely tropical family (Gut-

Fig. 5-7 Sawfly larvae *(Neodiprion rugifrons)* eating jack pine needles *(Pinus banksiana)*. After Knerer and Atwood (1973). Copyright © 1973 American Association for the Advancement of Science.

tiferae) that lacks furanocoumarins, but contains other light-activated toxins. Parallel evolution of light-activated toxins occurs in several plant families, and is met by adaptive radiation of resistant moths on each.

Existing patterns of herbivory show a variety of fascinating evolutionary phenomena. Some insects specialize on specific groups. Others use many unrelated hosts. Where specialization occurs, patterns of allelochemical defenses suggest an evolving defensive system that eliminates some insect herbivores, but fails to exclude others. However, there is no evidence that species formation or other taxonomic diversification of insects *causes* plant radiation, any more than pinworm evolution causes primate evolution.

Is it possible that the chemistries of different species of host plants cause insect speciation? Incipient speciation of insects on different hosts suggests that plants might force divergence in some insect populations. **Host races** on related plants are common in nature (Futuyma and Peterson, 1985). These are populations of insects of the same species, but with strongly evolved differences in host preferences. Host races are often considered incipient species that have not yet attained complete reproductive isolation from each other. An example is the pine sawfly *(Neodiprion abietis)*, which has highly developed genetic specialization on different conifer species (Fig. 5-7; see Knerer and Atwood, 1973). Presumably with time, reproductive isolation would evolve among sawfly forms because hybrids would be unable to digest any given conifer species as well as the sawfly populations adapted to that conifer. This kind of disruptive selection may be common in nature. Its importance in coevolution is that insect speciation may be in process as plants evolve. As plants diverge over evolutionary time, their many insect herbivores may diverge with them.

Polyphagous Herbivores and Plants

Coevolution between plants and polyphagous herbivores is diffuse. Silicification of grasses on the Serengeti Plains of East Africa provides a good example of local adaptation to herbivory from the plant perspective. S. J. McNaughton and J. L. Tarrants (1983) found that the grass *Eustachys paspaloides,* grown under controlled conditions, had a higher silica content if it was from heavily grazed ranges (3.9%) as compared with lightly grazed ranges (2.3%). Moreover, clipping (and presumably grazing in the field) increased silica content. *Eustachys* apparently protects itself against dozens of large grazing ungulates with both an inherited ability to respond to consistently high grazing pressure, and with inducible silicification. The response in this case is to grazing by any large mammal, not to a particular species of mammal.

SUMMARY

Genetic studies show that host specificity can be Mendelian or polygenic, and can affect either insect selection of food plants or larval survival on them.

Similarly, resistance to insects may be due to differences in alleles at one locus, to alleles at several loci with independent effects on different traits, or to polygenic effects on a single trait. The theoretical possibilities for different kinds of coevolution with such genetic complexity are immense.

Ecological variation and genetic complexities usually preclude the symmetrical, reciprocal adaptation of plant and herbivore that early theory predicted. Instead, a remarkable variety of evolutionary processes underlies plant and herbivore relationships. In the coarse-grained environments of sedentary herbivores and long-lived hosts, animal adaptation is much more rapid than plant adaptation. At the other extreme, food plants are fine-grained environments for browsers and grazers that eat dozens to thousands of plants in a lifetime. Plants may evolve defenses against especially voracious polyphagous herbivores much more rapidly than the animals adjust to the plants. It is unwise to view plant defenses in light of only one or two common contemporary herbivores. The evolution of plant defense occurs in a milieu of interactions with both the physical environment and numerous real or potential herbivores and disease organisms. Clearly the *co* must be added to *evolution* with care.

STUDY QUESTIONS

1. Distinguish between biotypes, host races, and species. What is the importance of each in plant and animal coevolution?
2. When might herbivore selection on a plant species be (a) stabilizing, (b) disruptive, and (c) directional? When might plant species impose these on an herbivore?
3. How might the evolution of aphids and their hosts differ from the evolution of buffalo and their food plants?
4. Why are single gene effects on resistance more easily documented than polygenic effects? Might this bias our opinions about the genetic basis of insect and host coevolution?
5. What information would you need to predict the frequency of cyanogenic clovers if rust fungi infect the cyanogenic forms, while slugs eat acyanogenic morphs?
6. The example of selection of dominant or recessive traits (Fig. 5-2) used a selection coefficient of 0.1. When in nature might the selection coefficient be much higher?

SUGGESTED READING

D. S. Falconer (1981) explores the principles of evolutionary genetics in the second edition of his classic work, *Introduction to Quantitative Genetics*. An anthology edited by Fowden Maxwell and Peter Jennings (1980), *Breeding Plants Resistant to Insects*, reviews the genetics of coevolution between agricultural plants and their pests. *Coevolution*, an anthology edited by Douglas Futuyma and Montgomery Slatkin (1983), provides an impressive variety of perspectives on many aspects of plant and animal coevolution.

III

MUTUALISMS:
THE UNEASY PARTNERSHIPS

The world would be a very different place without mutualisms between plants and animals. Pollinators make plant fertilization possible for most plant species by carrying pollen from one flower to another, and fruit-eating animals make seedling establishment possible for others by disseminating seeds. Innumerable tropical and many temperate zone plants use ants to defend themselves against herbivores or competing plants. If all insects, birds, and bats capable of transferring pollen suddenly vanished, most crops, wild herbs, trees, and shrubs would bear few fruits. Many would be entirely barren. If ants, mammals, and birds that habitually carry seeds suddenly vanished, seeds and seedlings would collect under parent plants and face almost certain destruction from pathogens, insects, and seed-eating rodents. Without protective ants, many plants would soon succumb to insect herbivores. Moreover, there is every reason to believe that the disappearance of key flowering and fruiting plants would push some dependent animals to the brink of extinction. Pollination, seed dispersal, and ant protection are only three of the mutualisms that make our world what it is. One of the exciting challenges of modern ecology is to acquire an understanding of the means by which plants and animals exploit each other to their mutual benefit.

6

Mechanics of Mutualism

Mutualisms exist when two or more species enhance each other's fitnesses. Symbiotic mutualisms involve species that live in close physical contact, like the gut microbes and herbivores discussed in earlier chapters. Of interest here are nonsymbiotic mutualisms between plants and animals in which species benefit each other, but do not live together.

Three processes that normally involve nonsymbiotic mutualisms between plants and animals are of special ecological importance (Table 6-1). **Pollination** is the transfer of pollen that results in the fertilization of egg cells in plant ovules. Most higher plants use nectar or excess pollen to lure insects, birds, or bats that forage at the flowers for food, and coincidentally transport pollen from one flower to another. **Seed dispersal** is the transport of seeds away from parent plants. Most species of seed-bearing plants use attractive fruits or the nutritious seed itself to attract birds, mammals, or ants that bury, regurgitate, or defecate viable seeds away from parent plants. **Protection** of plants by aggressive ants occurs when plants lure ants with sugar secretions or starch bodies on their stems, leaves, or buds. Some plants even provide hollow stems, thorns, or bulbs as shelter for ant colonies. The ants guard the source of sugar or starch, and simultaneously defend the plant against herbivorous insects or even en-

Table 6-1 Important mutualisms between plants and animals

Interaction	Advantage to Plants	Advantage to Animals
Pollination	Egg fertilization, especially by pollen from other individual plants	Nectar and/or pollen as food for insects, birds, and mammals; scents as mating pheromones for some (euglossine) bees
Seed dispersal	Escape of seeds and seedlings from pathogen, insect, and rodent attack near parent plants; dissemination to sites suitable for seedling growth	Food for insects or vertebrates that digest edible pulp and pass or regurgitate seeds, or hide seeds to eat and later fail to recover them
Protection by ants	Ants kill or discourage herbivores; ants clear vines and other competing vegetation away from host plants	Food from extrafloral nectaries, and starchy or protein-rich food bodies, shelter in hollow stems or thorns

croaching vegetation. Most terrestrial plant species require animals for pollination, seed dispersal, or protection. Some require all three.

Mutualisms often begin as antagonistic relationships between plants and animal enemies that eat their flowers or seeds (Thompson, 1982). Mutual benefit may evolve if plants can pay a tax in nectar, extra pollen, fruit pulp, or extra seeds to animals that help the plant reproduce successfully. For instance, a bee collects nectar and pollen on flowers and coincidentally brushes pollen from one flower onto the stigma of another flower of the same species. The plant uses a portion of its energy budget from photosynthesis to attract the bee with scent, colorful displays, nectar, and the fraction of its pollen that serves as bee food. These are reasonable sacrifices to bees if more pollen reaches the stigmas of potential mates, and if a plant's own stigmas receive more pollen from other flowers, than would be possible if pollen were transported by wind, water, or other kinds of flower visitors. From the vantage of natural selection, plant and animal mutualists inadvertently assist each other as they pursue their own reproductive self-interests.

POLLINATION

Angiosperm flowers attract pollinators with a rich variety of shapes, sizes, colors, and scents that serve as cues for a nectar or pollen reward. Likewise, insects and some birds and mammals are adapted in morphology, sensory capability, and digestive physiology to find and use that pollen and nectar.

Flower Function

The angiosperm **flower** evolved as an elaborate device for (1) capturing pollen from other plants of the same species and (2) dispatching its own pollen to other flowers of the same species. Flowers of animal-pollinated plants generally conform to a fundamental plan (Fig. 6-1). The ovule is the egg-containing structure that ultimately develops into the seed. One or more ovules is enclosed in a protective maternal structure, the carpel. The fused basal portion of one or more carpels forms the ovary which, when mature, becomes the fruit. One or more carpels with accessory parts for receiving pollen (stigma on a style) constitutes the pistil. A fleshy receptacle transfers nutrients from the rest of the plant to the other reproductive organs. Petals and sepals are modified leaves with two key functions: (1) they attract animals with bright pigments and sometimes with nectar-producing tissues and (2) they clasp and protect the developing flower bud or ovary from hungry insects. Protection was probably the original function of petals and sepals, with pollinator attraction a more recent evolutionary development.

The function of angiosperm pollination is the union of a haploid sperm nucleus with a haploid egg nucleus. the following events lead to fertilization: (1)

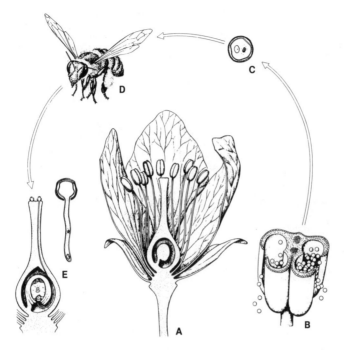

Fig. 6-1 Pollination of a simple angiosperm flower. The flower (A) consists of the pistil containing the ovary with the ovule(s), a ring of anther-bearing stamens, a ring of brightly colored petals called the corolla, and a ring of sepals called the calyx. A ripe anther (B: cross section) releases the pollen (C). An insect, here a honeybee, (D) carries pollen to the stigma of another flower of the same species. There the pollen grain germinates and a pollen tube (E) grows down to the ovary, where the egg contained in the ovule is fertilized. After Barth (1985).

a pollen grain reaches a stigma of a plant of the same species, (2) a tube cell develops from the pollen grain, extending as a pollen tube down the style to the nucellus, (3) one generative cell ($1N$) from the pollen grain fertilizes the egg cell ($1N$) to form a diploid zygote ($2N$), and (4) a second generative cell ($1N$) from the pollen grain unites with two polar nuclei (each $1N$) to form a triploid endosperm nucleus ($3N$). The diploid zygote becomes the embryo and eventually a juvenile plant, nourished by reserves supplied by the triploid endosperm. Together, the zygote (later the embryo), endosperm, and membranes of the former ovule comprise the **seed.** The fruit, defined earlier as the mature ovary containing seeds, sometimes also contains tissues derived from the receptacle.

Animals usually play an integral role in pollination, both from the perspective of pollen donation by flowers (male function) and pollen acceptance leading to seed development (female function). Most temperate angiosperm species are insect-pollinated. Even a prairie community, dominated by a half-dozen species of wind-pollinated grasses, may contain 250–300 species of insect-pollinated herbs and shrubs. Almost all tropical angiosperms are pollinated by

Fig. 6-2 Pollen transfer by animals. (A) A ruby-throated hummingbird *(Archilochus colubris)* probes for nectar in a trumpet creeper flower *(Campsis radicans)*. Pollen dusted onto the bird's bill or forehead may be brushed off on the stigma of the next flower visited. (B) A bumblebee *(Bombus)* packs clover *(Trifolium)* pollen into corbiculae on the rear tarsi for transport back to its nest. Pollen caught on the insect's body hairs may likely be brushed off onto stigmas of other clover flowers. (A) Photograph by Michael Hopiak, Cornell Laboratory of Ornithology; (B) photograph by Chip Clark, Smithsonian Institution.

insects, bats, or birds. Moreover, the reliance of plants on pollen transfer by animals has subtle consequences for plant biology. As will be seen in later chapters, the efficiency of pollen transfer by animals or other agents (male function) may have profound effects on the ways in which plants ensure proper fertilization and seed and fruit development (female function).

Pollination Syndromes

Natural selection by different animals with different sensory capabilities, nutritional needs, and habits has molded a remarkable variety of flower adaptations, ranging from superficial changes in color or scent to fundamental changes in the number and shape of flower parts (Fig. 6-2). **Pollination syndromes** are suites of flower colors, scents, and shapes that serve as cues used by insect or vertebrate pollinators to locate flowers (Table 6-2).

Ancient angiosperms were almost certainly **entomophilous,** or pollinated by insects that fed on pollen or the flower itself. Successive adaptations ensured attraction of potential pollinators in ways that minimized pollen loss and damage to plant reproductive parts and maximized pollen transfer.

B

Beetle pollination is probably the primitive condition in flowering plants. Many of the earliest flowers, like some modern orchids *(Listera cordata)*, probably emitted fetid odors that attracted beetles or other insects that normally ate rotting fungi, carrion, or dung. Contemporary beetle flowers, such as magnolias, usually have little or no nectar and attract visitors with enormous quantities of pollen, much of which is consumed by the insects. Only slightly later,

Table 6-2 Pollination syndromes that associate different taxonomic groups of flower visitors with different flower characteristics

Animal	Flower				
	Opening	Color	Odor	Shape	Nectar
Entomophilous					
Beetles	Day/night	Dull or white	Fruity or aminoid	Flat or bowl-shaped; radial symmetry	Often absent
Carrion or dung flies	Day/night	Brown or greenish	Fetid	Flat or deep; often traps; radial symmetry	Rich in amino acids, if present
Bee flies	Day/night	Variable	Variable	Moderately deep; radial symmetry	Hexose-rich
Bees	Day/night	Variable but not pure red	Sweet	Flat or broad tube; radial or bilateral symmetry	Sucrose-rich for long-tongued bees; hexose-rich for short-tongued bees
Hawkmoths	Night	White or pale green	Sweet	Deep, often with spur; radial symmetry	Ample and sucrose-rich
Butterflies	Day/night	Variable; often pink	Sweet	Deep or with spur; radial symmetry	Often sucrose-rich
Primarily Pollinated by Vertebrates					
Bats	Night	Drab, pale; often green	Musty	Flat "shaving brush" or deep tube; often on branch or trunk; hanging; much pollen; radial symmetry	Ample and hexose-rich
Birds	Day	Vivid; often red	None	Tube; often hanging; radial or bilateral symmetry	Ample and sucrose-rich

Source: Adapted from Howe and Westley (1986).

developments included flowers that were adapted for pollination by butterflies and moths. Tubular flowers, bright colors, copious nectar, and viscous pollen presently characterize flowers adapted to attract long-tongued butterflies (e.g., the tropical legume *Caesalpinia pulcherrima),* which were among the first insects adapted to sip nectar rather than eat pollen or flower parts. Strongly scented, night-blooming *Silene noctiflora* draws hawkmoths, the nocturnal equivalents of butterflies. Flower adaptations for attracting such tried-and-true pollinators are still common in a world now occupied by far more advanced plants and pollinators. Beetles, butterflies, and moths undoubtedly dominated communities of flowering plants 100 million years ago.

As the angiosperm revolution became firmly established in the earth's biota, plants evolved sophisticated means for luring increasingly diverse flower-visiting insects, birds, and bats (Fig. 6-2). Relatively recent adaptations include explosive anthers in "buzz-pollinated" flowers of the genus *Solanum* (potato family), which are triggered by vibrations from alighting insects (Buchmann, 1983), and fragrant but nectarless orchids like *Cypripedium* (Dressler, 1981). Orchids transfer pollen in enclosed pollen sacs called pollinia, which snap onto the backs of visiting bees. Tropical orchids attract male euglossine bees with chemical fragrances that the bees use to attract mates, while bees carry pollinia from one flower to another without even an opportunity to eat the pollen itself.

Plants have also evolved rewards that in some degree reflect the needs of their pollinators. Nectars vary widely in overall sugar concentration, ranging from 15–88% by weight. Within this range are distinctive categories of sugar ratios that are associated with different kinds of pollinators. For instance, hummingbirds strongly favor a high ratio of the disaccharide sucrose to the monosaccharides glucose and fructose, while sunbirds, honeycreepers, and other nectar-feeding birds, as well as many insects, favor much lower disaccharide to monosaccharide ratios (Table 6-3). Herbert and Irene Baker (1983) found that nectars also contain amino acids in concentrations high enough to be of nutritional importance to pollinators. Amino acids are in high concentrations (12.5 μmol/ml) in nectars that attract insects that do not eat pollen, such as carrion and dung flies, but are also appreciable in flowers that attract other insects (0.5–1.5 μmol/ml). Amino acid concentrations are lowest (<0.5 μmol/ml) in plants pollinated by bats and birds that feed heavily on pollen or insects.

Flower-visiting vertebrates and the plants adapted to attract them are rela-

Table 6-3 Number of plant species in four sugar-ratio categories arranged by principal pollinators

Pollinators[a]	Sucrose/(Glucose + Fructose) Ratio			
	<0.1	0.1–0.499	0.5–0.999	>0.999
Hummingbirds	0	18	45	77
Other birds	59	17	2	0
Hawkmoths	2	8	19	32
Other moths and butterflies	8	31	35	44
Short-tongued bees	115	103	28	17
Long-tongued bees	13	75	49	66
Bats	10	21	2	1
Beetles	1	3	2	3
Flies	29	27	7	9

Source: Data from Baker and Baker (1983).

[a]Related taxonomic categories not significantly different from each other are pooled (e.g., other birds).

tively recent in the fossil record. Tough, resilient flowers of many members of
the balsa family, such as the baobab tree *(Adansonia)* of East Africa, use a
musky smell to attract pollen-eating bats. The numerous stamens and anthers
of bat-flowers often give them the appearance of shaving brushes. Bats foraging
from one tree to another are dusted with pollen, which they brush off on subsequent flowers that they visit. This may appear to be a clumsy system of
pollen transfer. However, wide-ranging foraging flights of the bats ensure that
widely scattered plants are pollinated, and reduce inbreeding among plants and
their adult offspring nearby. Other flowers adapted to attract vertebrates deco-
rate both tropical and temperate communities with brilliant red tubular flowers
(Fig. 6-2A). In Africa and Asia such plants attract sunbirds, and in the Amer-
icas hummingbirds. These are only a few of the many examples of known
pollination syndromes. A modern forest or meadow is a museum of threads of
evolutionary descent, some extending back only a few million years, some over
100 million years.

The degree to which pollination syndromes are represented in a community
can be a valuable clue to the importance of different kinds of animals to local
pollination biology. Abundant red tubular flowers in mountains of Central and
South America testify to the importance of hummingbird pollination at high
elevations where insects are disadvantaged by cool temperatures. A larger proportion of green, white, and yellow flowers in tropical American lowlands signals the more important role of flower-visiting insects in a more equable climate.

Useful as they are in some contexts, syndromes are sometimes fallible for
good biological reasons. The match between pollinator and syndrome is often
loose. For example, it is not unusual to see a "bumble-bee flower" like *Delphinium nelsonii* pollinated by hummingbirds (Waser, 1983). The birds may
visit *Delphinium* when other food is scarce, or simply when it is convenient to
do so. Moreover, some plants apparently adapt to different visitors, both as
populations and as individuals. In the Rocky Mountains of the United States,
individuals of the scarlet gilia *(Ipomopsis aggregata)* are deep red early in the
season, when hummingbirds are abundant, and light pink later in the summer,
when hummingbirds migrate away and hawkmoths are common (Paige and
Whitham, 1985). The *Delphinium* fits the bee syndrome but may be pollinated
by birds, while scarlet gilia takes advantage of both bird and moth syndromes
at different times during the summer.

Pollinators

Morphological, physiological, and sensory attributes of flower-visiting animals
influence their roles as pollinators. The morphological attributes of flower visitors have obvious relevance to pollination. Long retractable tongues and long
bills allow sphinx moths and hummingbirds to use long tubular flowers (fig. 6-
2A). Bees are equipped with tarsal structures, called corbiculae, which they
use to carry pollen back to their nests (Fig. 6-2B). Pollination occurs by accident, from the bee's perspective, when pollen grains caught in their body hairs

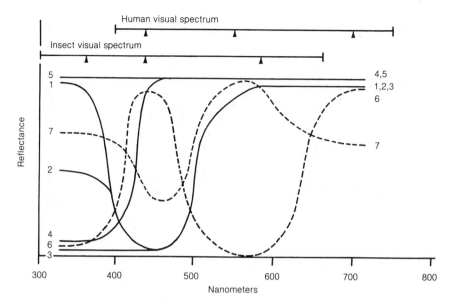

Fig. 6-3 Reflectance curves for human and insect visual spectra. Curve 1 is human yellow, insect purple; curve 2 is human yellow, insect reddish-purple; curve 3 is human yellow, insect red; curve 4 is human white, insect yellow; curve 5 is white for both; curve 6 is human purple, insect green; curve 7 is human greenish-yellow, insect mauve. After Kevan (1983).

are brushed off onto stigmas of the flowers that they visit. Flower-visiting bats have long, prehensile tongues that allow them to lap up pollen and nectar from the bottoms of deep tubular flowers. Like bees, bats become dusted with pollen in the process of eating flower products, and are consequently carriers of pollen to flowers on other trees.

Physiological characteristics of flower-visiting animals are less obvious than morphology, but no less important. Physiological necessity drives many of the most efficient pollinating agents. A 5-g hummingbird or sphinx moth uses 18 calories per minute while hovering in front of flowers (Wolf et al, 1973; Bartholomew and Casey, 1978). This roaring metabolism forces hummingbirds and hawkmoths to forage diligently on hundreds of flowers each day or night, making them good prospects as pollinators for plants capable of attracting them. Plants that consistently make use of these strong flyers offer copious sugar-rich nectar to entice them to visit, while plants that attract insects with much lower metabolic needs invest much less in nectar (Table 6-3). Animal metabolism molds flower traits, and flower distribution and characteristics determine which pollinators can flourish in a given woodland or meadow.

Finally, sensory capabilities differ widely among flower-visiting animals, and consequently shape the evolution of flowers used by different pollinators. Insects perceive different colors and shapes from vertebrates like ourselves. The compound eyes of bees are a receptive to a wider visual spectrum than the eye of a human (Fig. 6-3). Many flowers that appear a solid yellow or white to people are distinctly patterned to bees that perceive ultraviolet light, which

Golden cinquefoil   White bryony

Fig. 6-4 Ultraviolet nectar guides are invisible to humans, but are important visual cues to insects. To humans, golden cinquefoil is solid yellow and white bryony is solid white. Both are strongly bicolored to bees. After Barth (1985).

humans cannot see. Golden cinquefoil *(Potentilla aurea)* appears bright yellow to humans and bryony *(Bryonia dioica)* looks white. But to bees, dark ultraviolet flower centers contrast sharply with yellow and white, serving as **nectar guides** for directing the insects to the stamens and stigmas of the flower (Fig. 6-4). Different insect taxa are sensitive to different portions of the color spectrum, leaving plants the possibility of evolving a wide array of cues that are largely invisible to humans.

Insects also have a substantially different perception of shapes and smells from humans. Over 70 years ago, Karl von Frisch discovered that bees distinguish objects with different **figural intensity,** defined as the contour length or "edginess" of a figure. From a bee's perspective, the length of perceived contours implies high intensity. For instance, bees can distinguish a solid block from a hollow one, much as we can. But bees cannot distinguish a solid block from a solid triangle or a solid circle (Fig. 6-5). This means that flowers that appear to be very different shapes to us may be indistinguishable to bees. Insects are also sensitive to scents; bees detect sweet odors, flesh flies detect rotten protein. Not surprisingly, bee-pollinated flowers are fragrant, while those pollinated by carrion flies are rancid (Table 6-2). Flowers regularly pollinated by hummingbirds are scentless because hummingbirds have a poorly developed sense of smell. Differences in perceptual capabilities make one meadow several different landscapes to bees, flies, bats, and humans.

Fig. 6-5 Honeybees learn to discriminate the upper from the lower figures, but they cannot discriminate the upper figures from each other or the lower figures from each other. From Barth (1985).

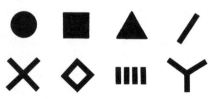

SEED DISPERSAL BY ANIMALS

Dispersal mutualisms are, if anything, older than pollination mutualisms. As early as 200 million years ago, the progenitors of modern cycads left fossils of fleshy seeds that were apparently adapted for consumption by primitive reptiles (Sporne, 1965). Like flowers, fruits clearly adapted to attract animals did not become common in the fossil record until the late Cretaceous, 65 million years ago. Since then the Earth has seen an incredible proliferation of ecological relationships between fruits and the birds, bats, monkeys, ungulates, and rodents that eat them (Howe, 1986). Almost absent in the harsh deserts of Africa and the Middle East, adaptations for animal dispersal reach their pinnacle in species-rich tropical rainforests, where up to 90% of the trees and virtually all shrubs bear fruits adapted to attract birds or mammals.

Fruit Function

An angiosperm fruit has three major functions: (1) nourishment, (2) protection, and (3) dispersal of plant embryos. Fruit structure is at least as varied as flower structure, but again is based on a fundamental plan common to all flowering plants (Fig. 6-6). Nourishment for the diploid zygote, or embryo, is provided by a triploid endosperm stocked with carbohydrates, fats, and protein by the parent plant. The embryo and endosperm are surrounded by a protective maternal diploid seed coat, to form the seed. A seed is a matured ovule. A **fruit** includes one or more seeds, the matured ovary, and often other maternal tissues such as the receptacle.

Dispersal structures may be derived from any of several tissues. In some flowering plants the seed coat contributes winglike structures for dispersal by wind, or produces a fleshy and nutritious outgrowth called the aril for attracting vertebrates, or the elaiosome for attracting ants. In most animal-dispersed plants,

Fig. 6-6 Fleshy angiosperm fruits. Left: Pome (apple). Right: Drupe (cherry).

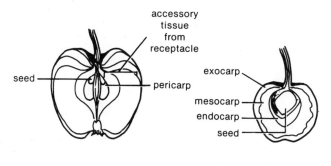

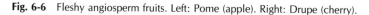

accessory
tissue
from
receptacle

seed

pericarp

exocarp

mesocarp

endocarp

seed

Pome (apple-an accessory fruit)   Drupe (cherry)

an edible pulp develops from the endocarp or mesocarp, both derived from the matured ovary, or exocarp, which is often derived from the receptacle. In many fruits these three layers are combined into the **pericarp,** which serves as the attractive lure to animals or the source of winglike structures for wind dispersal (Fig. 6-6). In the **drupe** of a cherry or plum, the outer mesocarp and exocarp of the pericarp are succulent, while the endocarp surrounds a stony seed. In a **berry,** such as that of a blueberry, many seeds are embedded in a fused pericarp.

Indehiscent fruits, such as drupes and berries, do not crack open; they mature as bitter and cryptic green versions of the ripe fruit. Dehiscent fruits are those with protective capsules that split open to expose a ripe, and often arillate, seed.

Table 6-4   Characteristics of fruits and seeds dispersed by different animals

| Animal | Fruit | | | |
	Color	Odor	Form	Reward
Primarily Dispersed by Vertebrates				
Hoarding mammals	Brown	Weak or aromatic	Indehiscent thick-walled nuts	Seed itself
Hoarding birds	Green or brown	None	Rounded seeds or nuts	Seed itself
Arboreal mammals	Yellow, white, green, or brown	Aromatic	Arillate seeds or drupes; often compound and dehiscent	Pulp protein, sugar, or starch
Bats	Pale yellow or green	Musky	Various; often hanging	Pulp lipid- or starch-rich
Terrestrial mammals	Often green or brown	None	Indehiscent nuts, pods, or capsules	Pulp lipid- or starch-rich
Highly frugivorous birds	Black, blue, red, green, or purple	None	Large drupes or arillate seeds; often dehiscent; seeds >10 mm long	Pulp lipid- or starch-rich
Partly frugivorous birds	Black, blue, red, orange, or white	None	Small or medium-sized drupes, arillate seeds, or berries; seeds <10 mm long	Pulp often sugar- or starch-rich
Feathers or fur	Undistinguished	None	Barbs, hooks, or sticky hairs	None
Primarily Dispersed by Insects				
Ants	Undistinguished	None to humans	Elaiosome on seed coat; seed <3 mm long	Oil or starch elaiosome with chemical attractant

Source: Adapted from Howe and Westley (1986).

Dispersal Syndromes

Dispersal syndromes are constellations of fruit colors, scents, shapes, and nutritional qualities that are associated with different means of seed dissemination by biotic and abiotic agents (Table 6-4). Analogous to pollination syndromes, seed dispersal syndromes represent adaptive responses to selection by animals that harvest and hoard seeds, swallow fruits and discard seeds, or passively carry fruits on fur or feathers.

Seeds regularly hoarded by rodents or birds often show little adaptation for dispersal, other than a thickened seed coat and rounded form. For instance, pinyon pines *(Pinus edulis)* and other bird-dispersed pines of the Western United States (e.g. *Pinus albicaulis, Pinus flexilis, Pinus monophylla)* have large rounded seeds that sit loosely in open cones (Fig. 6-7). Other pines have small winged seeds that are disseminated by wind. Pinyon seeds, acorns, and other nuts hoarded by rodents or birds are eaten by their dispersal agents. For these plants, effective dispersal occurs only when the animal loses a seed.

Some fruits dispersed by hoarders are more modified for dispersal. Examples are violet and other herb seeds, discussed in the introductory chapter, that possess specialized starchy structures (elaiosomes) that ants relish. Some tropical nuts have a pulp that partially satiates a rodent, and perhaps encourages burial rather than consumption of the seed. Elaiosomes and pulp represent an important step in the evolution of dispersal adaptations because they offer a seed-eating animal nutrition other than the seed itself. This reduces the conflicting interests of plants and the animals that might eat their reproductive structures.

The most highly developed modifications for dispersal are those adapted for consumption by fruit-eating birds and mammals (Fig. 6-8). Berrylike fruits of hawthorns *(Crataegus spp.)* appeal to migrating birds in the temperate autumn, and to nomadic waxwings *(Bombycilla)* or thrushes *(Turdus)* in the winter (Herrera, 1984a). Berries or berrylike drupes predominate among bird-dispersed plants in temperate regions, and are abundant throughout the tropics. Some tropical fruits are quite unlike anything found in temperate regions. The golfball-sized capsules of neotropical nutmegs *(Virola)* open to display a spectacular red arillate seed, which is eaten by toucans *(Ramphastos swainsonii)* and fruit-eating monkeys *(Ateles geofroyii)* (Howe, 1983). The nondescript greenish capsule of another common tree of Central and South America, *Tetragastris panamensis,* dehisces to expose a bright purple core, which offsets arillate, aromatic seeds that are much sought after by howling monkeys *(Alouatta palliata)*

Fig. 6-7 A rounded bird-dispersed seed of the pinyon pine *(Pinus edulis)* compared with the more typical winged seed of the wind-dispersed bristlecone pine *(P. aristata).* Drawing by M. Klein.

Pinyon   Bristlecone

Fig. 6-8 Fruits eaten by vertebrates. (A) Hawthorn *(Crataegus spp.)*, a favorite pome of wintering birds in temperate Europe and North America. (B) Wild nutmeg *(Virola surinamensis)*, an oily arillate fruit eaten by toucans in Central and South America. (C) Monkey fruit *(Tetragastris panamensis)*, a sugary fruit much sought after by monkeys and other arboreal mammals in Central and South America. Photographs by H. F. Howe.

and other arboreal mammals (Howe, 1980). Still other tropical fruits are dull but massive, like *Drypetes gossweileri* of dense West African forests, which is a favorite elephant food (see Fig. 9-10). These examples only hint at the enormous diversity of shapes, sizes, and colors of fruits adapted for animal consumption.

Table 6-5 Representative nutritional rewards of fruits eaten by birds and mammals[a]

Common Name (genus)	Percentage by Dry Weight			Dispersal Agents
	Protein	Lipid	Carbohydrate	
Temperate				
Cranberry *(Vaccinium)*	3	6	89	Birds
Hawthorn *(Crataegus)*	2	2	73	Birds
Pin cherry *(Prunus)*	8	3	84	Birds
Pokeberry *(Phytolacca)*	14	2	68	Birds
Strawberry *(Rubus)*	6	4	88	Birds
Tropical				
Bird palm *(Chamaedorea)*	14	16	55	Birds
Fig *(Ficus)*	7	4	79	Bats
Mistletoe *(Phoradendron)*	6	53	38	Birds
Monkey fruit *(Tetragastris)*	1	4	94	Monkeys
Wild nutmeg *(Virola)*	2	63	9	Birds

Source: Herrera (1984a); Howe and Vande Kerckhove (1981); White (1974).

[a]Note that temperate fruits are carbohydrate-rich and may or may not have substantial amounts of protein. Tropical fruits may also be lipid-rich. Values refer to aril or pulp content, excluding seeds that are not digested.

Fruits that are adapted to be eaten by birds and mammals frequently offer substantial nutritional rewards. The pulp is usually rich in carbohydrate, lipid, and/or protein (Table 6-5) in addition to water and indigestible fiber. The most common berries and drupes of temperate regions produce watery pulp that is low in protein (<5% by dry weight) and lipid (<5%) and high in sugars and starches (>50%); however, other fruits are high in protein (5–18%) or in lipids (10–63%). These differences appeal to the needs of different fruit-eating animals. Some plants offer quick energy in sugars to animals that normally eat protein-rich insects (51–68% protein; White, 1974). Other plants lure largely carnivorous or insectivorous animals capable of digesting oily high-energy rewards rich in lipids. Still other plants produce pulp rich enough in protein that it can, if eaten in quantity, fulfill the protein requirements of some birds and fruit-eating mammals, such as bats (Moermond and Denslow, 1985). This of course is possible because fruits are much easier to find in large quantities than insects or other animal prey.

Finally, some fruits stick to fur or feathers and are passively carried by animals (Sorensen, 1986). Fruits of the aster *(Bidens pilosa)* attach to bird feathers and may be brushed or pulled off a few yards from the parent plant, or thousands of miles away on an oceanic island. Such hitchhikers do not help the birds in any way, so the plants that produce them are not mutualists of birds. But as biogeographer Sherwin Carlquist (1974) has shown, chance long-distance seed dispersal by far-ranging birds accounts for much of the botanical diversity of tropical islands throughout the Pacific.

Dispersal syndromes have many of the same shortcomings as pollination syndromes. Plants are far more creative of structure, reward, and ornament than biologists are of categories, and fruit-eating animals often eat fruits that appear to be adapted for other kinds of animals. Nothing keeps an elephant from eating small fruits, if it so chooses! Furthermore, chance rather than consistent dispersal mutualisms explains some plant distributions. Peter Grant and his coworkers (1975) found that Galapagos owls *(Asio flammeus)* eat seed-crushing finches *(Geospiza)* with spurge seeds *(Chamaesyce amplexicaulis)* in their crops. The owls fly from island to island, scattering seeds as they discard the remains of the finches. *Chamaesyce* seeds cannot disperse across water without assistance because they sink. Flourishing populations of this spurge on several of the Galapagos Islands are best explained as the result of chance dispersal by owls rather than by mutualisms between plants and animals.

Dispersal syndromes do help biologists interpret local ecology. For example, most fruits in the rainforest of Barro Colorado Island, Panama are adapted for dispersal by birds, bats, or hoarding rodents, all of which are abundant at this site (Leigh et al., 1982). In contrast, a similar forest in Amazonian Peru has an exceptionally diverse primate fauna of 11 species of monkeys, and has a far higher proportion of yellow "monkey fruits" than Barro Colorado Island (Janson, 1983; Terborgh, 1983). In tropical forests with an even wider array of mammalian dispersal agents, such as the rainforests of West Africa, broadly overlapping syndromes can be identified for several categories, ranging from small, dehiscent monkey fruits to huge indehiscent elephant fruits (Gautier-Hion et al., 1985). Dispersal syndromes, like their counterparts in pollination biology, give a first approximation of the kinds of plant and animal mutualisms likely to prevail in a community.

Dispersal Agents

Animal attributes that promote seed dispersal are adaptations for efficient foraging and food digestion, not results of selection on animals to disseminate seeds. Even in the face of nearly a billion-fold range in the size of seed dispersers, from ants (0.01 g) to elephants (up to 7,500,000 g), only a few mostly tropical birds and monkeys differ appreciably from herbivorous or carnivorous relatives in digestive morphology and physiology (Fig. 6-9). Successful seed dispersal usually occurs when dispersal agents lose or fail to digest seeds intended as food.

Many rodents and birds hoard seeds for future use. These are seed predators because they digest and kill many or even most seeds that they collect. Coincidently, they are effective dispersal agents when they forget or lose buried seeds. Familiar mammalian examples are squirrels (e.g., *Sciurus* and *Tamiasciurus)* that collect several species of oak acorns *(Quercus)*, walnuts *(Juglans)*, and hickory nuts *(Carya)* (Smith and Follmer, 1972). Less familiar examples include tropical rodents such as the agouti *(Dasyprocta)*, as well as a host of species of mice and rats (e.g., *Mus, Peromyscus, Reithrodontomys)*. In all of these cases the primary adaptations for seed dissemination are behavioral tendencies to bury seeds singly or in small caches, sometimes abetted by cheek pouches that allow the animal to carry more than one seed at a time.

Seed-hoarding birds are at least as important as rodents in some parts of Eurasia and in montane communities of western North America. Crows, jays,

Fig. 6-9 Vertebrate dispersal agents. (A) Squirrels such as this chipmunk *(Tamias striatus)* hoard seeds for food. Some seeds are not recovered, and germinate. This chipmunk has filled its cheek pouches with seeds. (B) Like most large herbivorous mammals, cattle *(Bos taurus)* digest most seeds that they eat, but pass others unharmed. (C) Many fruit-eating birds like this cedar waxwing *(Bombycilla cedrorum)* regurgitate large seeds and pass small ones unharmed. (A) Photograph by L. Elliott, Cornell University; (B) photograph by H. F. Howe; (C) photograph by A. Cruikshank/VIREO.

Fig. 6-10 Nutcracker *(Nucifuga columbiana)* dispersal of pines. (A) Nutcracker collecting seeds from open cones of the single-leafed pinyon *(Pinus monophylla).* (B) A forgotten cache of whitebark pine *(Pinus albicaulis)* seeds has germinated. (A) Photograph by M. Bledsoe; (B) photograph by S. Vander Wall.

and nutcrackers, all members of the family Corvidae, disseminate vast numbers of pine seeds and acorns (Fig. 6-10). Stephen Vander Wall and Russell Balda (1977) found that Clark's nutcracker *(Nucifuga columbiana)* carries seeds of the pinyon pine *(Pinus edulis)* up to 22 km, burying them in small clumps that are ideal for germination and establishment. The nutcracker possesses a well-developed throat pouch (Fig. 6-11), analogous to a rodent cheek pouch, that allows it to carry 30 to 120 pine seeds at a time, depending on the species of pine and the size of its seeds (Bock et al., 1973). These common mountain birds of the western United States store two to three times as many seeds as they need to eat, apparently protecting themselves against theft by other seed-

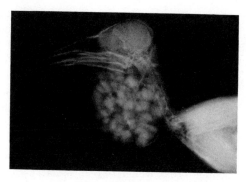

Fig. 6-11 Nutcrackers carry up to 120 seeds of the various bird-dispersed pines in a throat pouch. This pouch contained 38 large seeds of the single-leafed pinyon. Seeds may be cached hundreds or even thousands of meters from collecting sites. X-ray photograph by S. Vander Wall.

eating animals. Vander Wall and Balda estimate that a flock of 150 birds may store 3 to 5 million seeds in a season!

Other animals consume dry or fleshy fruits, and defecate some seeds in viable condition. For instance, horses *(Equus caballus)* eat a wide variety of fruits. These large herbivores digest 44 to 100% of the seeds that they eat, depending on the species (Janzen, 1981). Undigested seeds are defecated 14 to 60 days after they are eaten. This kind of digestive seed predation and seed dispersal is usually associated with large herbivorous mammals, such as elephants, antelopes, and cattle, and with large herbivorous birds, such as cassowaries and emus (Howe, 1986). However, smaller herbivores can play the same role. Mallard ducks *(Anas platyrynchos),* for instance, kill 70% of the seeds of the aquatic plant *Najas marina* that they eat, but pass 30% through the gut alive and ready to germinate (Agami and Waisel, 1986). Like rodents and nutcrackers, these herbivorous birds and mammals are both seed predators and dispersal agents. Unlike seed hoarders, they do not possess structural, physiological, sensory, or behavioral adaptations that enhance seed dissemination beyond their habit of consuming fruits and seeds. They simply fail to digest some seeds.

Fruit-eating birds and mammals that depend heavily on fleshy fruits for food are termed **frugivores.** Highly frugivorous animals, unlike nocturnal seed-eating rodents or large herbivorous mammals, often have color vision, making it possible for plants to attract them with visual displays as well as with scent. Perhaps most important are gut modifications that help frugivores process fruit quickly, while at the same time resulting in gentle treatment of seeds in the digestive tract.

Digestive morphology and physiology are best understood in light of the challenges faced by animals that digest only fruit pulp, as compared with those that digest foliage and seeds. Most plant-eating vertebrates have grinding teeth, stout bills, or pebble-filled muscular gizzards, as well as highly elaborate guts. Together, these adaptations permit plant tissues to be broken down and processed by the microbial symbionts that can digest fibrous tissues and detoxify poisonous secondary compounds. Such treatment is dangerous to all but the toughest seeds. Frugivores have traded fibrous and toxic food for a diet that is easy to find and digest, but is usually deficient in protein (Table 6-5). Fruit-eating birds and mammals must eat 2 g of fruit for each gram of body weight in order to extract enough protein for maintenance (Moermond and Denslow, 1985). To a 30-g summer tanager *(Piranga rubra)*, this requires the consumption of the equivalent of 200 6-mm-wide berries each day. Sometimes these extra calories are a burden for frugivores. The oilbird *(Steatornis caripensis)* of Trinidad feeds its young entirely on fruit (Snow, 1961). The young grow slowly in nesting caves, reaching a peak weight of 650 g—240 g *above* adult weight— in 70–80 days. Having eaten enough protein to develop adult muscles, these pudgy youngsters must fast away excess fat for a month before they can fly!

To meet the challenge of using a calorie-rich but protein-poor diet, fruit-eating birds and mammals either supplement their diets with protein-rich insects or minimize the time that they take to process fruits. Animals that supplement their diets of foliage or insects with fruits do not have distinctive digestive tracts. For instance, birds that eat both insects and fruits do not have obvious digestive modifications (Herrera, 1984b). They regurgitate or defecate seeds and use fruit pulp largely as a quick energy source.

Specialists that process large quantities of fruits have abbreviated digestive tracts that speed the passage of seeds, and ensure their gentle treatment in the gut. For example, the colon of the obligately fruit-eating spider monkey *(Ateles)* has only half of the surface area of the colon of a howler monkey *(Alouatta)*, which is a leaf-eater of the same size (Hladik, 1967). Likewise, fruit-eating birds often have short intestines and lack a gizzard, permitting rapid processing of fruits. The phainopepla *(Phainopepla nitens)*, a mistletoe specialist of the southwestern United States, can eat a mistletoe berry *(Phoradendron californicum)* and defecate the seed in as little as 12 minutes (Walsberg, 1975). Other birds regurgitate bulky seeds in even less time, allowing more room in the gut for processing fruit pulp. Unless fruits happen to be seasonally scarce, fruit specialists simply allow most fats and carbohydrates in their food to pass through the gut without absorption. Rapid processing and the need to consume large quantities of fruits to maintain protein balance play to the interests of plants, which benefit from rapid and gentle seed treatment and high rates of seed dissemination.

ANT/PLANT MUTUALISMS

Ant/plant mutualisms are common in the tropics, and not uncommon in temperate communities. Ants assist plants in many ways. Ants occasionally pollinate plants, and they are often dispersal agents for herb seeds that have starchy or oily elaiosomes (Chapter 1). Ants also protect plants from herbivores or encroaching vegetation. Ant-plants (those receiving protection) frequently provide food and shelter for protectors. Symbiosis is common between plants and protective ants.

Ant Attractants

Many plants lure ants with food and shelter. **Extrafloral nectaries** are nectar-secreting organs on leaves, stems, or occasionally on bracts outside the inflorescences. Extrafloral nectaries do not assist in pollination, but they do attract ants that attack and usually prey on other insects that land on the host plant. Some specialized ant-plants produce glycogen- or protein-rich **food nodules** that are harvested by ants, and replenished as needed. Such specialist plants usually also provide shelter in **domatia,** or living quarters. These may be swollen thorns or hollow stems, often with a weakened wall that allows ants to penetrate to the interior of the plant easily.

The most spectacular mutualisms between ants and plants are found in tropical forests and savannas. A famous example was first described over a century ago by Thomas Belt in *The Naturalist in Nicaragua* (1874). Belt noticed that aggressive ants occupied several Central American plants, and appeared to defend them from animals and from the encroachment of other vegetation. Daniel Janzen (1966) explored this and other apparent mutualisms more thoroughly,

Fig. 6-12 Bull-thorn acacia branches *(Acacia cornigera)* are occupied by an aggressive stinging ant *(Pseudomyrmex ferruginea)*. Ants live in the thorns and feed on starchy Beltian bodies and sugar-rich secretions from extrafloral nectaries. The ants clear vegetation around host trees and attack insect and mammalian herbivores that might eat the foliage. Drawing by R. Roseman.

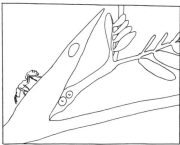

Fig. 6-13 Scanning electronmicrograph of a nearly mature extrafloral nectary of a passion flower *(Passiflora warmingii)*. An accumulation of nectar can be seen as a bulge beneath the cuticle. Photograph by L. T. Durkee. 285×.

discovering that some species of *Acacia* were in fact highly modified for ant occupation (Figure 6-12). For instance, *Acacia cornigera* is one of five "swollen-thorn acacias," characterized by inflated hollow thorns occupied by colonies of an aggressive stinging ant, *Pseudomyrmex ferruginea.* The *Acacia* maintains ants with nutritious nodules called **Beltian bodies** (named after Thomas Belt) on the ends of leaflets, and with extrafloral nectaries on leaf petioles. Supported by shelter, solid food, and nectar, the ants ferociously attack any herbivore that lands on their tree or vine intruding on their host's growing space. Similar complexity exists in the relationships of *Azteca* ants and *Cecropia* tree species that provide hollow stems and glycogen-rich food nodules called Müllerian bodies. Ants provide protection from leaf-eating herbivores and competing vegetation. These are only two famous examples of symbiotic mutualisms. Many tropical trees, shrubs, and epiphytes (plants living on the limbs and trunks of trees) are intimately involved with ant mutualists.

Ant protection is far more general than these well-known symbiotic mutualisms suggest. Many plants have well-developed nectaries on stems, stipules, or leaves, or they secrete sugar-rich liquid from unopened flower or leaf buds (Fig. 6-13). Such nectaries attract virtually any sugar-loving ants, most of which also eat other insects. In temperate climates, seedlings and developing buds often secrete nectar, and are defended by ants. For example, seedlings of cherry trees *(Prunus)* secrete nectar that encourages their defense by a variety of opportunistic ants (Tilman, 1978). Though by no means as diverse as flower and fruit modifications, adaptations for the attraction of ant mutualists are widespread and important.

Ants That Help Plants

Pollination and seed dispersal by fruit-eating vertebrates are conspicuous mutualisms, so it is ironic that some of the most specialized mutualists are inconspicuous ants that vigorously defend food plants. The reason for this specialization is that ants are often as dependent on their host plants as the plants are on their insect defenders. Any benefits to plants from pollinator activity or travels of dispersal agents are strictly incidental to the needs of the insect, bird, or mammal that visits the plant. There is nothing incidental about assaulting a caterpillar that is chewing on the acacia tree that provides a home, nectar, and Beltian bodies to ants. Mutualisms between acacias and their protectors really are of vital and immediate self-interest for both parties.

The principal adaptations of ants to plant defense are behavioral. Barbara Bentley (1976) found that normally carnivorous ants defend herbs on which she sprinkled sugar water. The opportunists ate other insects that they happened to encounter near their artificial nectar.

A far more organized defense characterizes the protection given by *Pseudomyrmex* and *Azteca* ant species that defend *Acacia* and *Cecropia* trees (Janzen, 1966). Both species sting and attack vegetation as well as insects or mammals that touch their hosts. *Pseudomyrmex* even clears vegetation around its host acacia. Even greater specialization is evident in the defense of west African *Barteria fistulosa* trees against large browsing herbivores. Unlike other plant-ants, *Pachysima aethiops* has a sting capable of penetrating elephant hide or numbing a lesser mammal for days (Janzen, 1972). Behavioral traits conducive to finding and stinging large mammals, but not small insects, leave no question that *Pachysima* is specially modified for *Barteria* defense.

SUMMARY

Plants and animals show a wide array of mutualisms in which two or more species benefit each other. The most common and conspicuous nonsymbiotic mutualisms involve two critical stages in plant reproduction: fertilization and seed dispersal. Flower size, shape, color, and pollen and nectar content are adjusted in innumerable ways to attract appropriate insects, birds, or mammals. Likewise, maternal tissues of the seed coat or ovary are modified as fruits of widely differing sizes, colors, and nutritional contents for attracting dispersal agents. Few of these nonsymbiotic relationships are limited to species pairs.

Most involve at least several animals that are compatible with a variety of similar flowers or fruits.

In contrast to pollination and seed dispersal, in which animal activity only coincidently benefits plant reproduction, ant defense mutualisms are sometimes highly specific. Certain ant species are specialized for defending particular plants that provide shelter and food bodies instead of, or in addition to, extrafloral nectar. These are among the most highly developed plant adaptations for defending themselves against herbivores other than through production of toxic secondary compounds.

STUDY QUESTIONS

1. What flower types would be most common in a hot lowland tropical forest? In a cool mountain forest?
2. Should natural selection favor a plant with fruits that meet all of the nutritional needs of fruit-eating birds? Why or why not?
3. Toxic secondary compounds are common in unripe fruits, but usually are absent from ripe pulp. Why?
4. In what ways are herbivory and seed dispersal mutualisms related?
5. What advantage might ants have over toxins for plant defense?

SUGGESTED READING

Friedrich G. Barth (1985) provides an elegant discussion of plant adaptations for pollination and of the perceptual qualities of insect pollinators in *Insects and Flowers*. The classic monograph on structural modifications of fruits is Leendert van der Pijl's (1972) *Principles of Dispersal in Higher Plants*. Barbara Bentley and Thomas Elias (1983) edit a useful anthology relevant to both floral ecology and ant protection of plants in *The Biology of Nectaries*.

7

Ecology of Mutualisms

Mutualisms depend on the ability of organisms to use each other as resources. Charles Darwin (1859; p. 62) notes the tendency of people of his time to "behold the face of nature bright with gladness" in a world of cheerful cooperation. Darwin understood the implications of natural selection for mutualisms. He contrasts the nineteenth century image of birds helping to disperse mistletoe berries with the observation that birds eat the fruits for their own purposes, and leave most mistletoe seeds in dense clumps where seedlings are doomed to intense competition with each other. If Darwin were to rewrite his passage today he might note that a mistletoe berry is largely indigestible seed and sticky glue that adheres to host plants. The plant shortchanges birds as much as it can, while still providing attractive enough fruits to fulfill a dispersal function. Darwin's skillful contrast of cheerful natural efficiency with the reality of a desperate struggle for survival among parasitic plants is an eloquent image of a classic seed-dispersal mutualism.

If "struggle," to use Darwin's term, underlies mutualism, why are mutualisms so widespread? What prevents each mutualist from exploiting its partners without providing a benefit? The answers lie in an evaluation of the degree to which relationships diminish or enhance the fitness of each mutualist. Fitness costs and benefits may be evaluated through **optimization theory,** first borrowed from economics (Maynard Smith, 1978). Optimization theory assumes that organisms balance benefits (advantages) and costs (disadvantages) of each activity or function so as to maximize genetic contributions to future generations. Explicit algebraic models can be used to predict how mutualisms diminish or enhance the fitness of each species involved (Chapter 8). This chapter explores the ecology of cost and benefit in mutualistic relationships.

POLLINATION

Pollination is a critical step in the sexual reproduction of flowering plants. Plants encourage animal foraging behaviors that result in dispersal of their pollen (male function), and in fertilization and seed development (female function). Pollinators favor plants that enhance their own foraging efficiencies. A

useful way to explore costs and benefits in pollination biology is to compare
the ideal consequences of pollinator activity from both the plant and animal
perspectives.

   The short-term interests of plants and pollinators are not the same. In the
best of all possible plant worlds, each egg cell is fertilized by a generative
sperm nucleus carrying alleles that would confer the highest possible fitness for
conditions that the seedling is likely to face in its local habitat. In reality,
foraging habits of insects, bats, or birds determine whether an ovule receives
genetically different pollen from other individual plants, genetically similar pollen from its own flowers, or any pollen at all. A bee cannot know or care
whether the pollen that it carries offers adaptive genetic combinations for flowers that it visits, since that value is determined by the success of seeds produced
long after the insect's death. In the best of animal worlds, a single visit to a
flower provides all of the pollen or nectar that an animal can use to sustain
itself, produce eggs, or feed to its young. A bee in paradise would only need
to find one flower. But the real world to a bee is a landscape of colors, shapes,
and scents that advertise nectar and pollen rewards that are less than the jackpot
that would serve all of a forager's biological needs. Plants provide enough
inducement to ensure visitation, but not so much that a pollinator can forgo
carrying pollen to other plants.

Foraging Behavior

Pollinators choose among flowers of different individuals and species as they
collect pollen or nectar. Each aspect of foraging behavior presents intriguing
questions. Does a bee or butterfly visit flowers of one species or several? How
does it find flowers with abundant pollen and nectar? Can it avoid flowers
already depleted by other insects? As critical, from the plant perspective, is
what the pollinator does *after* collecting pollen. Does a bee leave one apple
blossom for another on a neighbor, stop to eat pollen, return to its hive, or fly
to a field of sunflowers?

Cost and Efficiency

Bees are among the best-studied pollinators, and consequently offer revealing
insights into the costs and benefits of flower visitation (Heinrich, 1979a; Barth,
1985). The first lesson is that bee lifestyles require an enormous expenditure of
energy. Bees forage for nectar and pollen so that they may feed and often
incubate larvae. The larvae may be housed in large communal hives (e.g., the
honeybee *Apis mellifera),* in small colonies (e.g., large bumblebees, *Bombus*
spp.), or in brood chambers occupied by a single female and her young (e.g.,
the small adrenid bees, *Adrena* spp.). Many bees warm their young and forage
during cold weather, showing considerable thermoregulatory capacity. This energy expenditure by a small animal requires an immense foraging effort. A
bumblebee weighing 0.5 g may use 150 calories an hour, alternating brood

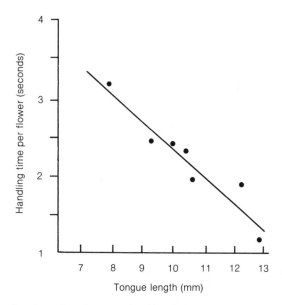

Fig. 7-1 Tongue lengths and working speeds of bumblebee species visiting *Delphinium barbeyi*. Data from Inouye (1977).

tending with foraging for the equivalent of 500 blueberry flowers worth of nectar. There is little leeway for wasteful foraging. Constraints are implicit in an energy-intensive lifestyle (Heinrich and Raven, 1972). A bee cannot waste time or effort on flowers deficient in nectar or pollen, or fumble around with flowers that are difficult to use. In theoretical terms, a bee cannot allow energetic cost to exceed benefit.

Bumblebees illustrate how handling efficiency influences foraging behavior. Bumblebee species with short tongues take nearly three times as long to handle tubular larkspur *(Delphinium barbeyi)* flowers as bumblebees with tongues only 60% longer (Fig. 7-1). Handling delays cut time available for visiting other flowers. Different species of pollinators sort to those flower types that they can most efficiently use. Bernd Heinrich (1976b) provides an example of flower partitioning by bumblebees. Long-tongued species dominate flowers with deep corollas; short-tongued species dominate either flowers with shallow corollas or flowers that provide only pollen, not nectar (Fig. 7-2). Bee species adjust their foraging behavior to the flower resources that they are best equipped to use.

Foraging and Competition

Foraging behavior should evolve in such a way that pollinators can adjust to a dynamic landscape of flower resources. Flexibility is important because flowers of different species secrete nectar at different times during the day and night. In addition, flowers of most species are visited by several to many species of

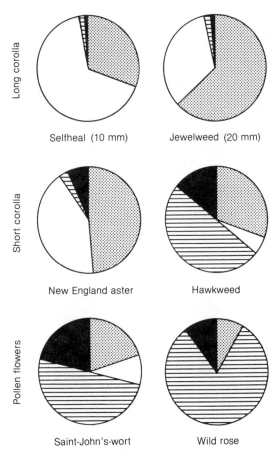

Fig. 7-2 Bumblebee partitioning of long-corolla flowers (corolla depth in parentheses), short-corolla flowers (corolla < 3 mm long), and pollen flowers in Maine. The four bumblebee species are *Bombus vagans* (stipples), *B. fervidus* (open), *B. terricola* (lined), and *B. ternarius* (black). Note that long-tongued bumblebees (*B. vagans* and *B. fervidus;* tongue > 8 mm) dominate flowers with long corollas, while short-tongued species (*B. ternarius* and *B. terricola;* tongue < 7 mm) predominate at flowers with short corollas and at pollen flowers. After Heinrich (1979b).

insects (Fig. 7-3), resulting in the frequent depletion of nectar or pollen by competitors.

Heinrich (1979b) found that bumblebees in Maine systematically **major** on flower species with high rewards and **minor** on alternative flower species. Major flower species are those most profitable at a given time, while minor flower species are fallback food sources. Young *Bombus vagans* are indiscriminate on their first trips out of the hive, foraging on unrewarding as well as rewarding flowers. After just a few trips, the bees concentrate on the most rewarding flowers (e.g., jewelweed *Impatiens biflora)* but continue to sporadically visit other flowers (Fig. 7-4A). Experimentally induced competition forces the bees

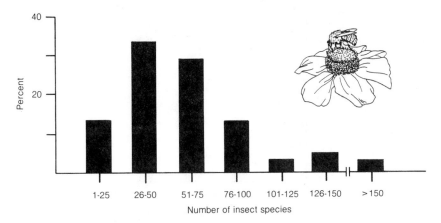

Fig. 7-3 Frequency distribution of insect vistation to 55 species of sunflowers (Asteraceae) in Illinois. Flowers of some plant species are visited by dozens to over 100 species of pollinators. Depletion of pollen and nectar by competitors creates a dynamic landscape of resources for insects. After Schemske (1983).

to widen foraging activities from one to as many as seven flower species, with an average of three to four species visited per trip (Fig. 7-4B). In an uncompetitive environment, the bees become extremely proficient at handling the odd-shaped jewelweed flower (Fig. 7-5), only occasionally diverting their attention to other flowers. Bees clearly profit from specializing to the point of proficiency on an abundant flower resource, but spread their efforts among a variety of flowering species when competitors deplete preferred flowers of nectar and pollen.

Fig. 7-4 Influence of experience on the foraging of a bumblebee *(Bombus vagans)*. (A) Decline in the number of flower kinds visited on consecutive foraging trips when there is no competition from other bees. (B) Constant number of flower kinds visited on consecutive foraging trips when competing bees deplete favorite ("major") flowers. After Heinrich (1979a).

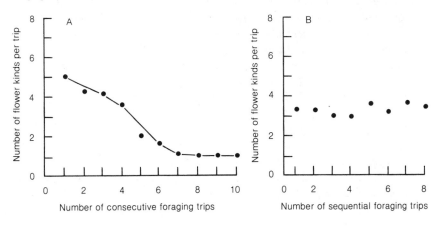

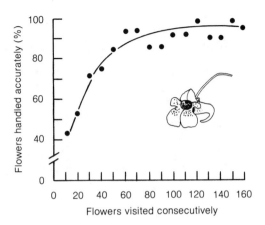

Fig. 7-5 Improvement of handling accuracy at jewelweed *(Impatiens)* flowers from the first to the 150th flower of a handling career. After Heinrich (1979a).

Competition produces different foraging responses in other pollinator species. Individual *Bombus fervidus* workers do not live as long as the flowering seasons of their food plants. Individuals from the same colony major on different food species throughout their lives. Under competition, individual bees extend their foraging ranges rather than switch to minor foods (Heinrich, 1976). Other pollinators solve the problem of competition with **territoriality,** in which they defend rich food resources against competitors. Large territorial *Trigona silvestriana* in tropical America find rich food sources and recruit swarms of sisters from the same hive. Hundreds or even thousands of biting stingless bees defend rich flowering trees far better than one or two individuals. These bees will often fight to the death in a furious defense of an exceptionally rich find (Johnson and Hubbell, 1974). Hawaiian honeycreepers *(Vestiaria coccinea)* defend the red-flowered tree *Metrosideros collinia* only when it has intermediate levels of flower production (Carpenter and MacMillen, 1976). Below a lower threshold of about 60 flowers, the energetic cost of fighting is too high for the return in calories from nectar. Above an upper threshold of 250 flowers, nectar is superabundant and there are too many intruders for these birds to mount an economical defense. Life history attributes, flower defensibility, and the presence or absence of aggressive behavioral repertoires determine which alternatives to flower switching are used in these pollinators.

How do pollinators find rewarding flowers, and avoid barren ones? Direct assessments and systematic foraging methods tend to keep bees in productive patches of flowers (Waddington, 1983). A bee may visit a flower briefly, leaving immediately if there is little nectar or pollen to be collected. Some bees detect recent visitors by scent, thereby eliminating the necessity to even alight on a flower. Behavioral devices also keep bees foraging in rewarding patches of flowers. The bumblebee *Bombus americanorum* flies very short distances

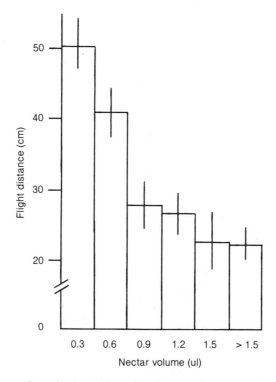

Fig. 7-6 Distances flown by bumblebees *(Bombus americanorum)* after visiting inflorescences of *Delphinium virescens* containing different amounts of nectar. Distances are means
accompanied by standard error bars. Nectar volumes refer to the mean volume of the lowest
two flowers on an inflorescence. After Waddington (1981).

after visiting larkspur *(Delphinium virescens)* flowers containing large quantities of nectar, and much longer distances after visiting flowers of the same
species with little nectar (Fig. 7-6). Once they find a good patch, bees take
advantage of prevailing conditions by returning again and again to the most
profitable part of a meadow or field.

Bees and wasps also perceive the variability of flower resources. Leslie Real
(1981) found that individual bumblebees *(Bombus sandersoni)* and paper wasps
(Vespula vulgaris) quickly associated the color of artificial flowers with the
constancy of sugar water that was experimentally added to them. These species
preferred artificial flowers of a color signaling a low variance in reward over
those with colors signaling extremely high rewards in some flowers and none
in others. Foraging for even a small but predictable reward apparently allowed
the bees to avoid exhausting themselves in consecutive visits to barren flowers.
Refinements of theory and experimentation by Real and his collaborators (e.g.,
Real and Caraco, 1986) have shown that a wide variety of foragers balance the
mean and variance of reward content in making foraging decisions.

Evolved Specialization

Pollinators may visit flowers of only one, or those of dozens, of species of plants. **Monolectic** pollinators visit flowers of just one *taxon* of plant, **oligolectic** pollinators visit several taxa, and **polylectic** pollinators visit many different taxa. The terms can apply to different taxonomic levels. An entomologist might refer to bees that are monolectic or polylectic to plant species, genus, or family. A bee polylectic to species might be monolectic to genus.

The vast majority of pollinator species are oligolectic or polylectic. This is inevitable for many pollinators that live longer than the flowering seasons of their food plants, and likely for any insect, bird, or bat faced with dozens or hundreds of possible flowers to visit. The immediate energetic advantages in foraging efficiency of specializing on one or a few flower resources are outweighed by such sources of ecological variability as competition for preferred species and local and seasonal differences in plant species abundance.

If most flower visitors are continually assessing the costs and benefits of using different species, genera, and even families of plants, why do some insects feed on only one *species* of plant? Monolecty to species would appear to be unprofitable except when a short-lived insect uses a common long-lived flower resource. Two examples of monolecty illustrate the unusual circumstances that favor use of a single flower resource.

Some insects raise their larvae on the ovules or seeds of the same plant species that they pollinate. A classic example is that of 600 species of figs *(Ficus)*, each pollinated by its own unique species of minute wasp (family Agaonidae; Ramirez, 1974). The unusual natural history of figs makes this possible. Fig flowers are embedded in an enclosed receptacle, called the **syconium** (Fig. 7-7). A ripe syconium attracts tiny female wasps by scent. A female wasp, already mated and carrying pollen, enters a syconium through a pore, dusts pollen onto some fig flowers while laying eggs on others, and dies. Each larva consumes the ovule in which it hatches. Some larvae mature into wingless males that mate and die, while others become females that mate, chew their way out of the matured fig, and fly to a ripe syconium on another tree. Each mature fig fruit contains hundreds to thousands of seeds, approximately half of which are alive, half killed by wasp larvae. Another classic example involves seed predators *(Tegeticula* moths) that, like fig wasps, are also pollinators of their food plants *(Yucca* plants; Aker, 1982; Addicott, 1986). Unlike most pollinator/plant relationships, which are nonsymbiotic, fig wasps and yucca moths are both symbiotic mutualists and seed predators.

Other plants may consistently present an abundant flower resource for a short time, making extreme specialization profitable. Some short-lived bees restrict virtually all of their feeding activities to one or at most a few species of short-lived plants. A classic example is the solitary bee *Adrena rozeni*, which feeds almost exclusively on the annual primrose *Camissonia clavaeformis* of the southwestern deserts of the United States (Linsley et al., 1963, 1964). The bee begins foraging late in the afternoon, just as the small flowers of this short-lived desert plant open. The pollen grains of the primrose are on long sticky

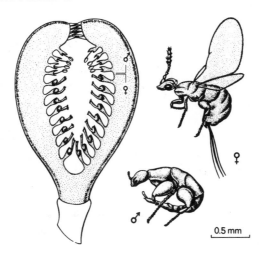

Fig. 7-7 A fig *(Ficus)* syconium and male and female fig wasps *(Blastophaga)*. Female wasps enter the syconium and lay eggs on some ovules and pollinate others. Wingless males mature, mate, and die. Winged females burrow out of the mature fig, carrying pollen to a ripening syconium. From Barth (1985); also see Galil (1977).

strings that require a certain degree of behavioral specialization to handle. Spring flowering in the desert comes and goes so quickly that many bees, and in particular *Adrena rozeni,* find it easiest to become extremely proficient in using one species, rather than taking the time to adjust to many. *Camissonia clavaeformis* is partly dependent on *Adrena rozeni* because the little bee is so abundant. But the primrose is also visited by other bees. A similar example involving a specialist bee *(Hoplitis anthocopoides)* and a superabundant but ephemeral flower *(Echium vulgare)* (Strickler, 1979) in the eastern United States indicates that sparse desert habitats are not prerequisites for obligate relationships between plants and insects.

Few well-documented examples of monolecty exist, suggesting that obligate plant and pollinator relationships require special circumstances. Whether these unusual cases involve paying a tax in ovules rather than nectar and pollen, or such explosive flowering that only one or a few pollinators can make use of the flowering season of a plant species, closely knit mutualisms between pairs of species are special cases in pollination, rather than exemplars of general phenomena.

Flower Success

The success of a pollination mutualism depends on two related factors: foraging behavior of pollinators and the influence of that behavior on the donation of pollen and fertilization of ovules. Maximum genetic fitness in pollination is an abstraction because it would require that pollen and ovules be genetically matched for maximum seed and seedling fitness. For present purposes, flower success will be defined by fertilization success rather than genetic success.

Determining what *limits* fitness is an important aspect of pollination biology. The procedure for exploring limitations is to compare actual pollen and ovule success with theoretical success in the "best of all possible worlds" for both pollen and ovules.

Limiting factors for plant reproduction can be pollen donors or ovules to receive them, or environmental influences. When pollen generative nuclei fail to fertilize eggs, seed production is ovule limited. If some ovules are not fertilized, seed production is said to be pollen limited. Seed production is resource limited if eggs are fertilized, but aborted because there is not enough light to provision the endosperm or because soil nutrients available to the plant are insufficient for fruit development (Stephenson, 1981). Seed production is herbivore limited if ovules are fertilized but are later eaten by animals before maturity (see Chapter 10).

The proportion of flowers that produce fruits (number of fruits/number of flowers) is termed **fruit set;** the proportion of ovules that produce seeds (number of seeds/number of ovules) is termed **seed set.** Fruit and seed set are not synonymous. Fruit set measures the extent to which flowers set at least one seed, but it does not measure the number of seeds per fruit. For a plant with many ovules per flower, a 100% fruit set might hide a very low seed set if many ovules are never fertilized, are aborted, or are destroyed by insect larvae.

Plant reproduction is frequently ovule or pollen limited due to unreliable pollination. The chance that any given pollen grain will find a proper stigma in a wind-pollinated species is miniscule. Wind-pollinated plants must produce prodigious quantities of pollen—up to one million pollen grains for every ovule—to ensure reproduction through pollen donation. Even much more reliable insects rarely approach perfect service from the plant perspective. Most pollen grains are eaten by foragers, fed to larvae, deposited on inappropriate stigmas, or simply brushed off on vegetation. Some are never carried away. From the perspective of male function, reproduction is virtually always ovule limited. Even so, pollen might still not be limiting if all eggs in all ovules are fertilized.

How does one know whether pollen is limiting? Paulette Bierzychudek (1982) gives one clear example in the jack-in-the-pulpit *(Arisaema triphyllum)* in New York. Flower number is correlated with plant size in this species, but seed set is not, unless flowers are hand-pollinated (Fig. 7-8). Jack-in-the-pulpit plants are capable of maturing far more seeds than they do in a natural setting. Fungus gnats, the usual pollinators, are not common enough in this forest to result in a high seed set. Pollen limitation in jack-in-the-pulpit highlights the chasm between the real and best of all possible worlds for jack-in-the-pulpit reproduction. In other plants, herbivores, seed predators, or light or nutrient limitation might limit seed production.

Many plants adapt to pollination constraints through adjustments of their ability to fertilize themselves. Most plant species are **monoecious,** producing both male and female parts on the same individual. Many of these are **self-compatible,** meaning that they are capable of fertilizing their own egg nuclei

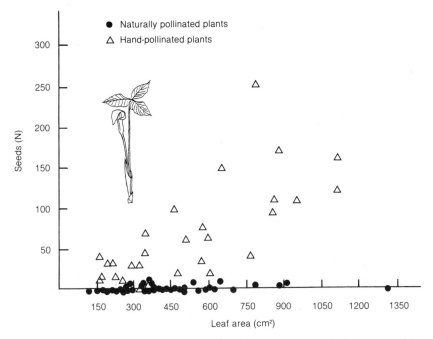

Fig. 7-8 Pollinator limitation in the jack-in-the-pulpit *(Arisaema triphyllum)* in New York. Seed production is strongly correlated (r = .81, p < .001) with leaf area in hand-pollinated plants, but not in individuals naturally pollinated by fungus gnats. After Bierzychudek (1982).

and are to varying degrees genetically inbred. Pollination success is likely in self-compatible plants if pollen from other individuals is in limited supply. **Self-incompatible** species are those in which individuals cannot fertilize themselves. For them, pollination often fails. The extreme plant sexual system occurs in **dioecious** species, in which sexes are separated on different individuals. Dioecism is almost universal among animals, but it is rare in herbaceous plants (1–9% of species) and is uncommon among trees and shrubs (11–33% of species; Bawa, 1980a). Dioecious species are, of course, outcrossed.

The extent to which self-pollination occurs influences pollination success. Pollination is certain for those plant species capable of pollinating themselves without visits by animals, fairly likely for those capable of selfing when a pollinator deposits a plant's pollen on its own stigma, and less likely if pollen from other individuals is necessary. Robert Cruden (1977) has found that the ratio of pollen to ovules produced by monoecious plants closely reflects the likelihood that flowers achieve pollination (Table 7-1). Plants that always pollinate themselves produce as few as 1 or 2 pollen grains per ovule, while insect-pollinated species that *must* outcross may have 10,000 pollen grains per ovule. Intermediate ratios reflect intermediate dependence of selfing as compared with outcrossing.

Pollen/ovule ratios even vary within species in which pollination success is

Table 7-1 Pollen/ovule ratios of plants with different
breeding systems

Breeding System (no. of species)	Pollen/Ovule Ratio (mean ± standard error)
Always selfs with flower closed (6)	5 ± 1
Always selfs with flower open (7)	28 ± 3
Usually selfs (20)	168 ± 22
Usually outcrosses (38)	797 ± 88
Always outcrosses (25)	5859 ± 936

Source: After Cruden (1977).

variable from one population to another (Cruden, 1976). In the tropical legume *Caesalpinia pulcherrima,* the proportion of hermaphroditic flowers ranges from 11 to 88% in different populations, corresponding to pollen/ovule ratios ranging from 9,416 to 827. High pollen/ovule ratios occur when large numbers of staminate flowers hold large numbers of pollen grains per anther. Heavy investment in male function (high pollen/ovule ratio) occurs in highland forests where nectar-feeding butterflies are scarce. Low pollinator activity favors individual plants that dust rare visitors with immense quantities of pollen. In forests where butterflies are common, individual plants that invest heavily in ovule and seed production are favored by natural selection.

Flower success depends on the success of both male and female parts of the flower (Willson, 1983). The different pollen/ovule ratios among and within species indicate that natural selection acts differently on male and female function. Genetic markers allow a direct assessment of pollen as well as ovule success. Norman Ellstrand (1984) has found multiple paternity in up to 85% of the wild radish *(Raphanus sativus)* fruits surveyed. Electrophoretic analysis of protein variants, corresponding to allele variants, implicates a minimum of two to four fathers in fruits having only two to eight seeds each. Maureen Stanton and her colleagues (1986) have further explored pollen and ovule success in another radish, *Raphanus raphanistrum.* Pollinating bees and butterflies strongly discriminate between two petal color morphs of this species. Less visited white and heavily visited yellow flowers have equivalent *maternal* success, as judged by seed set. But the heavily visited yellow flowers are far more successful as pollen donors than are white flowers (Fig. 7-9). The *paternal* success of yellow morphs is higher than that of white morphs. Put another way, there are enough pollen grains to go around for all available *Raphanus raphanistrum* ovules, but flowers compete for the opportunity to donate pollen. Plants with flowers that are capable of attracting the most pollinators fertilize egg cells before plants that have less effective flowers have a chance.

Differences in individual success through male and female function in pollen-limited trumpet creepers *(Campsis radicans)* suggest that sexual conflicts may be widespread in monoecious plants (Bertin, 1985).

Potential conflicts between male and female reproductive functions are strongest when pollinators obtain a resource from pistillate flowers that they cannot ob-

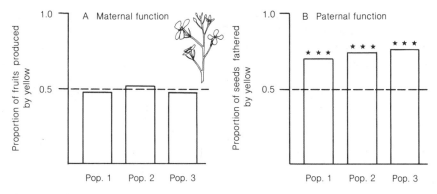

Fig. 7-9 Relative performance of yellow as compared with white morphs of the wild radish *(Raphanus raphanistrum)*. Pollinators favor the yellow morphs. This does not influence the extent of which yellow flowers produce seeds (A) because neither yellow nor white flowers are pollen limited. However pollinator preference for yellow does influence the extent to which yellow-flowered plants reproduce as pollen donors (B). Stars indicate that a significantly higher proportion of seeds are fathered by yellow than white flowers. After Stanton et al. (1986).

tain from staminate flowers (Lloyd and Bawa, 1984). For instance, bees and bee-flies feed on pollen, which is found only in staminate flowers, as well as nectar. This imposes a "cost of sex" on plants because bees avoid female flowers (Bierzychudek, 1987). For example in sarsaparilla *(Aralia hispida)*, flower clusters (umbels) alternate between male and female phases. James Thomson and his colleagues (1982) found that bumblebees favor male-phase flowers that offer both pollen and nectar, while bees tend to ignore female-phase flowers that offer only nectar. Asymmetry in bee preferences has molded flowering schedules. Umbels are male-phase for 4–5 days, relying on regular visits by bees that consistently patrol the same plants (trapline). Umbels are often female-phase for only a day, perhaps relying on bees that mistakenly visit flowers at which they usually collect pollen.

Sexual conflicts are most evident in dioecious species (Bawa, 1983). Bumblebees and bee-flies (e.g., *Syrphus ribesii;* Syrphidae) so strongly favor male Swedish brambles *(Rubus chamaemorus)* that fruit set is sometimes pollen-limited (Ågren et al., 1986). Pistillate flowers compete with staminate flowers that are inherently more attractive to pollen-feeding insects. Pistillate flowers resemble staminate flowers and secure pollination when insects that visit nearby male plants mistakenly search them for pollen. The extent to which pistillate flowers mimic staminate flowers is even more extreme in some tropical species. For instance, the stigmatic surfaces of pistillate flowers of the tropical tree *Jacaratia dolichaula* closely mimic petals of staminate flowers (Bawa, 1980b)! In this case female plants compete for hawkmoth pollinators in search of nectar produced, paradoxically, only by males.

In summary, pollinator influence on plant gender allocation is an exciting frontier in the ecology of plant and animal relationships. Genetic markers demonstrate that pollinator activity can profoundly influence the relative success of male and female function, even when each flower has both male and female parts. Different pollen/ovule ratios among and within plant species imply a dynamic adaptation to the ease with which plants reproduce through female and male functions. Bee and fly pollinators, in particular, favor pollen-producing staminate flowers. Females of dioecious plants must often compete directly with pollen-producing males, securing successful pollination by virtue of their resemblance to males, proximity to males, pollinator error, or by infrequent but effective visitation by bees seeking nectar. A wealth of intriguing questions remain to be asked in this vast field.

SEED DISPERSAL

Seed-dispersal and pollination mutualisms offer intriguing analogies and contrasts (Table 7-2). Both are nonsymbiotic mutualisms that depend on animal foraging behavior and plant nutritional rewards. In both mutualisms, plants risk the destruction of reproductive tissues by animal visitors. Pollination and seed dispersal differ in the degree to which plants influence the outcome of pollen transfer and seed dissemination. Adaptations for pollination ensure visits by animals carrying the appropriate pollen. Fruiting plants attract animals that eat fruit, but the plants cannot influence the destination of seeds that are consumed. While it benefits a plant to have a pollinator transport its pollen to flowers of the same species, it does not benefit a plant to have its seeds taken to another plant of the same species. Seeds or seedlings that accumulate under a neighboring tree of the same species are probably as likely to be killed by insects, rodents, disease, or competition as those that fall under their own parents. Moreover, few pollinators digest flowers, but many frugivores digest seeds. The inability of plants to guide dispersal agents to particular seed "targets"

Table 7-2 Contrasts between pollen and seed dispersal from the plant perspective

	Pollen Dispersal	Seed Dispersal
Target	Stigma of the same species, sometimes or always on another individual	Any place suitable for establishment
Animal motivation to target	Collect fragrances, nectar, or pollen	None: seeds are ballast to be discarded
Cues to target	Flower color, shape, fragrance	None
Advantage to visitor constancy	High: ensures pollen transfer	High or low: depends on behavior after leaving plant

Source: After Howe and Westley (1986).

and the destructive habits of many frugivores favor generalized fruit adaptations (Wheelwright and Orians, 1982). Generalized adaptations attract a variety of potential dispersal agents, reducing the likelihood that plants come to rely on specialized dispersers that evolve into destructive seed predators.

With the exception of ants, most animals that disperse seeds are vertebrates that are capable of carrying seeds long distances. Most pollinators are insects or small vertebrates that transport pollen grains a few centimeters to meters from their sources. Ants, like pollinators, are small (e.g., 0.01 g) and in fact bury seeds within a few centimeters of the plant from which they were collected. Birds and mammals that disperse seeds are virtually all larger than pollinators (>10 g), and many are very large (5,000–7,500,000 g; Howe, 1986). Large size has important implications for the kinds and amounts of nutritional rewards that fruits must offer to attract dispersal agents. Large body size also ensures that seed dispersal by animals—unlike pollen dispersal by animals—is usually measured in tens, hundreds, or thousands of meters rather than in centimeters.

Foraging for Fruits and Seeds

Animals with different feeding ecologies eat different fruits, and different fruit parts. Fruit- and seed-digesting finches, deer, grouse, parrots, pigs, horses, and elephants metabolize both pulp and seed constituents, and often digest even fruit capsules or pods. These herbivores digest the carbohydrate- or lipid-rich pulp as well as the protein-rich seeds of all but the most toxic or mechanically tough fruits. Rodents, jays, and other seed gatherers often ignore fruits with pulp, and store seeds for later consumption. Their challenges are to find, carry, and hide enough seeds to carry them through seasonal periods of scarcity and to compensate for losses to other animals that might find their caches. Frugivores such as fruit bats, thrushes, toucans, most monkeys, and many ants digest only fruit pulp. Their challenges are to find, eat, and subsist partly or entirely on foods that are usually deficient in protein, but are rich in carbohydrates or lipids.

Fruit Size and Abundance

All foragers must spend time and energy finding foods that vary in digestibility, nutritional quality, quantity, and accessibility. Optimization models help sort out the relationships of key variables influencing fruit choice by animals. Thomas Martin (1985) offers a simple cost/benefit model that illustrates constraints on frugivory by birds of different sizes.

In theory, the extent to which a bird profits from eating a particular fruit depends on the size of the bird, the size of the fruit, and the nutritional quality of the edible pulp or aril discounted by the effort expended acquiring it (Fig. 7-10). The model illustrates why different fruits have different profitabilities for birds of the same size, and why birds of different sizes prefer fruits of

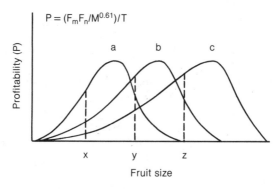

Fig. 7-10   Profitability (P) as a function of fruit size for birds of 17 (curve a), 250 (curve b), or 500 g (curve c). Handling costs of large fruits limit both the mean fruit size x and the width of the distribution of profitabilities (curve a) accessible to small foragers. Curves for larger birds are skewed because larger foragers find fruits profitable at larger mean sizes (Y for curve b, Z for curve c), but can often find abundant smaller fruits profitable as well. $M^{0.61}$ is the basal metabolic rate. After Martin (1985).

different sizes. For birds of the same size, profitability *(P)* depends on the pulp weight (F_m) and nutritional content (F_n) divided by the exploitation time *(T)*, which estimates the ease with which the birds obtain the fruits. Exploitation time depends on the abundance of fruits in a tree, the ease with which a bird finds a tree, and the ease with which fruits are handled once they are found. For small birds, exploitation time is prohibitively high for large fruits that are difficult to handle. Consequently, the profitability of large fruits for small birds is very low. For large birds, exploitation time is small for either large or small fruits that are common, but is prohibitively large for small fruits that are rare. One would not expect to find large birds searching for small, rare fruits. The profitability of fruits depends as much on bird attributes and the distribution of fruits as on the qualities of the fruits themselves.

Observations of frugivory by birds bear out the prediction that the ease of using fruits influences their dispersal. Nathaniel Wheelwright (1985) found that large birds eat large fruits and common small fruits, while small birds are limited to small fruits (Fig. 7-11). Also as expected, plant species with small fruits are often visited by more species of dispersal agents, and succeed in securing better dispersal, than plants with large fruits. In Central America the number of bird species visiting tree species is inversely related to fruit size (Table 7-3). This suggests that small fruits attract a wider array of dispersal agents than large ones, an inference borne out by direct observation in tropical communities (Martin, 1985). The relative advantage of small fruit size is also borne out in temperate communities. In the Mediterranean scrub of southern Spain, Carlos Herrera (1984c) found that shrub species with small seeds are depleted more rapidly than those with large seeds. Small fruit-eating birds (10–30 g) that are common in Spanish scrublands strongly prefer small fruits. Herrera (1981) points out that plants with large seeds must invest much more in

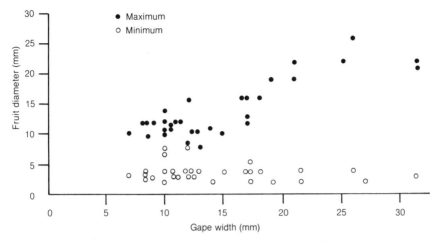

Fig. 7-11   Relationship between fruit diameter and gape (mouth) width for 36 bird species in the mountains of Costa Rica. Maximum diameters of fruit species included in a diet are positively correlated with gape width, and therefore with bird size. Large birds eat fruits inaccessible to smaller birds. Both large and small birds eat small fruits. After Wheelwright (1985).

nutritional reward to achieve the same rates of dispersal as plants with smaller, more easily handled fruits.

Fruit-eating animals may discriminate among the fruits of a single plant species. Some wild nutmeg *(Virola surinamensis)* trees disseminate much higher proportions of their fruit crops than others in Central Panama (Howe, 1983). Fruits of this large tree consist of a fatty aril (60% lipid) surrounding a large (1.5 × 2.0 cm) seed (Fig. 6-8). Toucans and other large birds prefer individual nutmeg trees with small rather than large fruits, and further favor trees bearing fruits that offer a high ratio of edible aril to indigestible seed (Fig. 7-12). Toucans are sensitive to both the costs of handling large fruits and the benefits

Table 7-3   Numbers of birds in feeding assemblages at Central American fruiting trees

Plant Species	Fruit Size (cm³)	Bird Species (N)
Virola surinamensis	2.7	7
Tetragastris panamensis[a]	1.8	12
Virola sebifera	0.7	6
Casearia corymbosa	0.5	22
Guarea glabra	0.2	19
Didymopanax morototoni	0.1	37
Miconia argentea	0.1	46

Source: Data from Martin (1985).

[a] Primarily a monkey fruit.

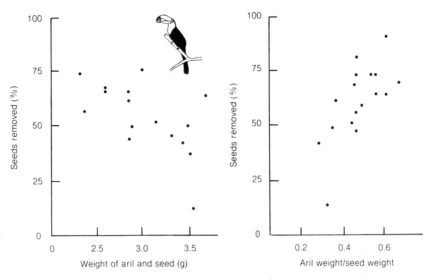

Fig. 7-12 Fruit removal from individual nutmeg *(Virola surinamensis)* trees by toucans and other birds in Panama. Left: The proportion of fruits removed declines with mean weight of the seed plus aril $(r = -.67, p < .002)$. Right: The proportion of seeds removed increases with the ratio of edible aril to indigestible seed attributable to each tree $(r = .72, p < .001)$. (A) After Howe and Vande Kerckhove (1981); (B) after Howe and Vande Kerckhove (1980).

of maximizing the nutritional reward for whatever cost in ballast or wasted gut space must be incurred by eating seeds.

Fruit accessibility and size often influence fruit profitability for potential dispersal agents. Timothy Moermond and Julie Denslow (1983) found that caged Central American birds show marked preferences among fruits. Some bird species, however, choose nonpreferred fruits that are easily collected over preferred fruits that are hard to pluck (Fig. 7-13). For instance, the tanager *Euphonia gouldi* prefers *Urera caracasana* over *Miconia affinis* berries when both are presented at perch level. The tanagers abandon this preference if *Urera* berries are placed below the perch so that the birds need to stretch to reach them. Frustrated tanagers simply switch to nonpreferred *Miconia* berries at perch level. Manakins *(Manacus candei)* do not switch as easily, continuing to eat *Urera* berries even when they must stretch or fly to pluck them. Clearly, fruit-eating birds differ in the extent to which accessibility overrides food preferences. Further experiments by Douglas Levey (1987) have shown that birds' handling techniques influence the sizes of fruits eaten. Manakins, which swallow fruits whole, are much more constrained by seed and fruit sizes than tanagers, which strip off fruit pulp and drop seeds.

In summary, complex interactions of nutritional reward, size, abundance, and accessibility of fruits influence the profitability of the fruits to dispersal agents. The most important single interaction is that between fruit and frugivore size. Fruit size helps and constrains seed dispersal. Plants that produce many small fruits attract the largest number and diversity of foragers (Snow, 1971).

Fig. 7-13 Fruit choice in experiments with caged birds. Tropical tanagers and manakins preferred some fruits to others, but for tanagers these preferences were easily overridden if nonpreferred berries were more accessible (perch level) than normally preferred berries (under the perch). From Moermond and Denslow (1983).

This sets up an inherent conflict for plants between the need to encourage dispersal of seeds and the need to provision seeds adequately for early seedling life. Large well-provisioned seeds in large fruits have an advantage in germination and establishment once they are disseminated. But they are not easily disseminated because they are difficult for all but the largest frugivores to eat.

Frugivore Specialization

Frugivores rarely specialize on only one or a very few species of fruiting plants. However, some birds do have a strong preference for certain fruit morphologies. For instance, paradise birds of New Guinea strongly prefer plant species with encapsulated fruits, while Asian fruit pigeons in the same rainforests prefer unprotected berries and drupes (Pratt and Stiles, 1985). But frugivorous vertebrates live longer than the fruiting seasons of most plants, and fruits of most plant species offer inadequate nutrition for a balanced diet. For both reasons, frugivorous birds and mammals forage widely for fruits of many species, as well as for other food, throughout their lives.

Sometimes relationships between fruit-eating animals and one or a few plant species appear to be tightly coevolved. These are usually local foraging responses to an accidental local food abundance. For example, fruit bats *(Arti-*

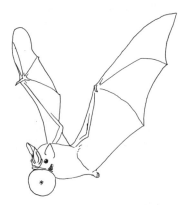

Fig. 7-14   Fruit bats *(Artibeus jamaicensis)* strongly prefer figs (here *Ficus insipida*) on Barro Colorado Island. An historical accident of extensive clearing makes figs, which are colonizing species, exceptionally abundant and predictable food resources on the island. As old fig trees fail to replace themselves on the island, which is now heavily forested, bats should shift to alternative foods. Elsewhere these bats eat a wide variety of fruits. Drawing by R. Roseman.

beus jamaicensis) strongly prefer two species of figs *(Ficus yopenensis* and *Ficus insipida)* to all other fruits eaten by mammals on Barro Colorado Island, Panama (Morrison, 1978). Figs amount to 78% of the fruits recovered from bat fecal samples (Fig. 7-14; Bonaccorso, 1979). In other forests these bats eat a wide variety of fruits (Heithaus et al., 1975). The best explanation for the apparent specialization on Barro Colorado Island is that figs happen to be exceptionally abundant and predictable food resources every month of the year. Like bees visiting artificial flowers, wild bats prefer a predictable food. The abundance of figs on the island is an historical accident. Figs colonized land that was cleared of trees during the building of the Panama Canal nearly a century ago. As aging figs die and fail to replace themselves in the lush forest now covering the island, the bats may be expected to switch to other foods.

A few rare instances of frugivore and plant coevolution exist. No well-documented example duplicates the level of evolved specialization seen in figs and their wasp pollinators, but lesser degrees of specialization occur among Asian flowerpeckers of the genus *Dicaeus* and several genera of parasitic mistletoes (Loranthaceae; Docters van Leeuwen, 1954). For some mistletoes, flowerpeckers are both pollinators and dispersal agents. In this case the association between disperser and plant is likely to be an evolved relationship, even though flowerpeckers eat fruits other than mistletoes, and mistletoes are dispersed by birds other than flowerpeckers.

Consequences of Seed Dispersal

Seed dispersal may enhance parental plant fitness in several ways. Seed dissemination may allow parent plants to establish their offspring in vacant sites, a

Table 7-4 Advantages to seed dissemination, from the perspective of parental fitness

Hypothesis	Advantage	Expected Attributes
Colonization	Occupy vacant sites	Small, often minute seeds; seed dormancy; rapid growth in sun; shade intolerant
Escape	Avoid density-dependent seed or seedling death near parent	Medium or large seeds; shade tolerant
Directed dispersal	Occupy specific micro-sites critical for establishment	Small, often ant-dispersed seeds; special germination or seedling requirements

process termed **colonization.** Alternatively, dispersal may represent **escape** from disproportionate seed and seedling mortality near the parent. In rare cases, animals consistently deliver seeds to special sites necessary for germination and establishment. This is termed **directed dispersal.** The advantages of seed dispersal for any given plant species can be formalized as hypotheses subject to field tests (Table 7-4). How might these hypotheses be distinguished?

Comparative natural history helps distinguish plants that benefit primarily by colonization from those that benefit primarily from escape. Colonizing plants are usually small-seeded species that germinate quickly if they are dispersed to open habitats, or remain dormant in the soil for years or even decades if they land in heavily vegetated sites. Often stores of dormant seeds, called **seed banks,** reach 200–3,800 seeds of various species per square meter of surface soil (Cook, 1980). Dormant seeds wait until a habitat opening comes to them. They germinate when soil is warmed by the sun following clearing, fire, or treefall, and seedlings quickly occupy the vacant space. Seeds that benefit primarily from escape are often large (>1 g) and usually do not show dormancy. Aggregations of large seeds are frequently infested with insects, or attract seed-eating mammals. Seed predation places a premium on dispersal that results in widely scattered seeds and seedlings that are difficult for insects and seeds to locate.

Colonizing plants often have long-lasting impacts on disturbed habitats. Animal-dispersed plants are often the first wave of occupants in the ecological succession of land cleared by fire, windfall, landslide, or human activity. Forestry operations in the mountains of the eastern United States often produce virtually solid stands of the bird-dispersed pin cherry *(Prunus pensylvanica),* which can remain dormant in the soil under a mantle of beeches, maples, and basswoods for over 50 years (Marks, 1974). Bat-dispersed figs, discussed earlier, include dozens of species that wait for years in seed banks in the soil until just the right conditions allow them to germinate and grow. The current abundance of large figs and the bats that service them on Barro Colorado Island is an ecological echo of past disturbance and colonization.

Directed dispersal is implicated when plants need distinctive habitats that represent a very small proportion of the total available habitat. The example of

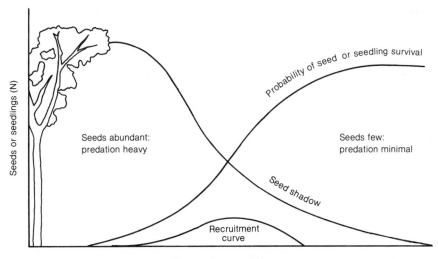

Fig. 7-15 Hypothesized escape of seeds from disproportionate mortality near parent trees. Seeds fall in a skewed distribution with the peak density under or near the parent. Mortality of seeds, seedlings, or saplings is greater near the parent than further away because of attacks by insects, pathogens, or rodents, or because of sibling competition. Consequently, overall survival increases with distance from the fruiting tree. A "recruitment curve" represents the distance at which greatest sapling survival is expected to occur. First proposed by D. H. Janzen (1970).

Asian mistletoes (family Loranthaceae) mentioned above is a good one. These parasitic plants germinate and establish only on the bark of host plants. Flowerpeckers eat the berries and rub strings of seeds off onto bark when they defecate (Docters van Leeuwen, 1954). These parasitic plants could not establish on host bark without flowerpecker assistance. Many ant-dispersed herbs probably benefit from directed dispersal. For instance, Diana W. Davidson and S. R. Morton (1981) found that ant-dispersed shrubs of two closely related genera, *Sclerolaena* and *Dissocarpus,* grow far more vigorously on ant mounds than elsewhere in the desert scrub of western Australia. Species not dispersed by ants also grow on the mounds, as well as elsewhere, but derive no special advantage from associations with ants.

The advantage to seed dispersal for other plant species is escape from disproportionate mortality in the immediate vicinity of parent plants (Fig. 7-15; Janzen, 1970; Connell, 1971). Parental fitness is enhanced by dispersal away from any source of density-dependent mortality near the parent, including seed-eating rodents or insect seed predators, infectious pathogens, and seedling or sapling competition (Howe and Smallwood, 1982). An example from tropical America illustrates how important escape can be for seeds of some animal-dispersed plants.

The American nutmeg *(Virola surinamensis)* discussed earlier bears an at-

Table 7-5 Proportion of fruits removed and dropped by frugivores at wild nutmeg (*V. surinamensis*) trees in central Panama[a]

Frugivore (no. sightings) (binomial)	Eaten/Visit ($\bar{x}$)	Of All Fruits Handled	
		Dropped (%)	Removed (%)
Black-crested guan (10) (*Penelope purpurascens*)	5	0	9
Massena trogon (81) (*Trogon massena*)	1	0	10
Rufous motmot (97) (*Baryphthengus martii*)	1	2	14
Collared aracari (16) (*Pteroglossus torquatus*)	<1	0	1
Chestnut-mandibled toucan (119) (*Ramphastos swainsonii*)	2	2	35
Keel-billed toucan (112) (*Ramphastos sulfuratus*)	<1	2	8
Masked tityra (40) (*Tityra semifasciata*)	0	5	0
Spider monkey (29) (*Ateles geofroyii*)	1	9	3

Source: From Howe and Vande Kerckhove (1981).

[a]Summarized from eight 5-hour watches at each of eight trees.

tractive arillate fruit (Fig. 6-8; Howe, 1983; Howe et al., 1985). In Panama, six species of birds commonly eat these fruits and either defecate (guans) or regurgitate (most others) viable seeds in the surrounding forest (Table 7-5). Other animals are less efficient for the plant. The tityra, too small to swallow this seed, simply picks off the aril and drops the seed under the tree canopy. Spider monkeys drop more seeds under the crown than they disperse. In some years, capuchin monkeys (*Cebus capuchinus*) feed heavily on the arils and drop seeds under the tree crown, or eat and digest the seeds themselves. The proportion of fruits taken differs among *Virola* trees (Fig. 7-12), but many always drop under the crowns. What is the effect of this frugivory on seed and seedling survival? This central question requires answers to three subsidiary questions.

1. Where do seeds go? Each year animals remove 46–60% of the fruits available, but for plants averaging 5,000 fruits each year upwards of 2,000 fall or are knocked down directly underneath the average parent. The net result is that seeds occur in densities of 120 per 10 m² directly under the crown, but a hundredth of that density only 20 m away. An average of half of the nutmeg seeds produced drop directly under the crown of their parent tree, while half are widely scattered in low densities throughout the surrounding forest.

2. What kills seeds? Rodents and deer eat some *Virola surinamensis* seeds, but small weevils of the genus *Conotrachelus* lay eggs on the vast majority of germinating seeds and young seedlings during the first 12 weeks after fruit drop. Weevil larvae virtually always kill the embryo (Fig. 3-1). Weevil depredations are much more severe under the tree crowns than only a few meters

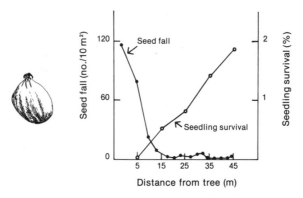

Fig. 7-16   Actual patterns of seed fall and seed and seedling survival three months after fruit drop in the toucan dispersed tree, *Virola surinamensis,* in Panama. Seed and seedling mortality due to *Conotrachelus* weevil infestations are so heavy under the crown that the seed fall and seedling survival curves cross at the crown edge, 10–15 m from the tree base. Not surprisingly, adults of this species are clumped, with an average nearest neighbor distance of 18 m. Overall, there is a 40-fold advantage to seed dispersal only 45 m from fruiting *Virola* trees. Data from Howe et al. (1985).

away (Fig. 7-16). Seeds dropped only 45 m from fruiting *Virola* trees are more than 40 times as likely to survive as those left under the crown.

3. Are all dispersal agents equally effective from the plant perspective? Even birds that transport and regurgitate viable seeds have strikingly different *indirect* effects on seedling survival. Smaller visitors such as trogons *(Trogon massena;* 145 g) and motmots *(Baryphthengus martii;* 185 g) drop up to 91% of the seeds that they consume within 20 m of the tree crowns, while much larger guans *(Penelope purpurascens;* 2,000 g) and toucans *(Ramphastos swainsonii;* 640 g) take most seeds more than 45 m. Because guans and toucans are much more likely to take seeds away from weevil infestations than smaller birds, they are much better dispersal agents, for every seed carried, than smaller species. Chestnut-mandibled toucans are 30 times as beneficial as rufous motmots for each seed carried. Because toucans eat more than twice as many seeds, the toucans are actually at least 60 times (2 × 30) as effective as motmots for *Virola surinamensis* dispersal.

Seed dispersal for any given plant species may be advantageous for more than one reason. There are good biological reasons for this; colonization, escape, and directed dispersal hypotheses are not exclusive (Howe and Smallwood, 1982). A plant could conceivably benefit from all three, and most plants probably derive some benefit from at least two. For example, the early advantage to *Virola surinamensis* seed dispersal is escape from weevil attack under the parents. But beyond 12 weeks after fruit fall animals no longer kill more seedlings under than away from fruiting trees. Shade-tolerant seedlings established in the forest understory wait for treefalls to open a sunlit path to the canopy. Widely scattered saplings are more likely to encounter one or more treefall gaps than survivors clustered under a long-lived parent tree. In short, an early escape advantage is superseded by a colonization advantage.

ANT PROTECTION

Many herbs, shrubs, and trees are guarded by ants that patrol their foliage or reproductive tissues in defense of extrafloral nectaries or other nutritious rewards. Once considered exotic oddities, protective ant/plant mutualisms are now known to be of widespread ecological significance.

Obligate Interactions

Protective ant mutualisms have had a long and quarrelsome history. Some prominent biologists doubted that ants actually defended plants, thinking instead that parasitic ants occupied *Cecropia* and *Acacia* trees to the detriment of the plants (Wheeler, 1942). Critics noted, for instance, that *Cecropia* trees sometimes lacked ants, and that even plants occupied by ants were vulnerable to herbivorous insects and sloths. Early evolutionists expected perfection in adaption. Variation in nature made adaptation suspect.

The dynamic nature of ant protection is now better understood. The beneficial role of ant protection to obligate ant-plants, such as *Acacia* and *Cecropia* species, is now beyond question. For example, Eugene Schupp (1986) thoroughly explored the seasonal dynamics of *Azteca constructor* protection of *Cecropia obtusifolia* in Ecuador. He found that *Cecropia* trees occupied by ants grew far more rapidly than those denied ants (Fig. 7-17). This advantage was

Fig. 7-17 Growth of *Cecropia obtusifolia* saplings with and without colonies of *Azteca constructor* ants in Ecuador. Ants effectively defend plants against beetles and other chewing insects and against vine competition. By the end of the first year, occupied plants are 2 m taller than unoccupied trees. After Schupp (1986).

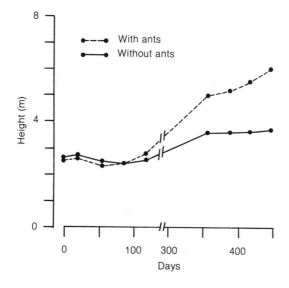

primarily due to ant protection from beetles. The ants were not especially effective against leafhoppers and flies. Schupp also found that ant protection was seasonal. Ants helped greatly in the dry season, far less during wet months. Schupp discovered, in a classic ant–plant mutualism thought to be well understood, that ant defense was inherently imperfect and variable, but was nonetheless critical to *Cecropia* growth.

Loss of Obligate Mutualisms

Protective ant mutualisms could be considered as appropriate in a chapter on herbivory as one on mutualism. They occur, like spines or toxic secondary compounds, because plants are forced by herbivores to invest in defense. Plants that invest in chemical defense often do not have ant defense, while relatives with inducements for ants often lack extensive chemical defenses. For instance, *Acacia* species that lack ant domatia, nectaries, or food bodies are well-defended with cyanide-producing compounds that do not occur in ant acacias (Rehr et al., 1973). A loss of ant mutualism places a premium on other forms of defense.

Other protective mutualisms are obligate early in the life of a plant, but dispensable later on. In West Africa, the ant-plant *Leonardoxa africana* produces nectaries on leaves and swollen stem internodes that are excavated and occupied by the ant *Petalomyrmex phylax*. Doyle McKey (1984) found that young shoots of this plant could not mature unless they were occupied by *Petalomyrmex*. Unoccupied shoots, or shoots occupied by an unaggressive ant *Cataulacus mckeyi*, were eaten by herbivores. Older *Leonardoxa* shoots and leaves had few nectaries and were not patrolled by ants. Old leaves suffered little herbivore damage because they were well-defended with tannins and lignins.

Facultative Protection

The opportunistic use of foliar nectaries by ants is much more common than such obligate interactions as *Cecropia* and *Azteca* mutualisms. The potential for opportunistic (facultative) ant–plant mutualisms is most obvious when a plant species with extrafloral nectaries is transplanted without the ants with which it has coevolved. In one example, extrafloral nectaries of European vetches *(Vicia sativa)* introduced to California are tended by Argentine ants *(Iridomyrmex humilis)* (Koptur, 1979). In another example, the Javanese ant-plant *Clerodendrum fallax* has been introduced to the Cape Verde Islands off the west coast of Africa without its Asian ants. A cosmopolitan ant normally not associated with plants, *Pheidole megacephala,* has invaded the hollow *Clerodendrum* stems (Jolivet, 1985). Both *Vicia* in California and *Clerodendrum* in the Cape Verde Islands are now actively defended by ants with which they share no evolutionary history. If even carnivorous ants can be induced to defend

Fig. 7-18 Inflorescence of *Costus woodsonii* defended by *Camponotus planatus* ants. The fly (*Euxesta* sp.), a seed predator (left center), has just been chased from active nectaries by ants. After Schemske (1980).

plants sprinkled with sugar water (Bentley, 1976), why are most plants not ant-plants?

Ant protection often involves several opportunistic ant species, each vying for dominance on resources offered by defended plants. For example, large tropical herbs of the genera *Costus* and *Calathea* produce large inflorescences studded with nectaries outside the flowers (Fig. 7-18). Ants patrol these inflorescences, chasing off insects that might lay eggs on or eat developing ovules and seeds. Two ants species repel a fly of the genus *Euxesta* that lays eggs on ovules of *Costus woodsonii* along the coasts of Panama (Schemske, 1980). Effective defense by the ant *Wasmannia auropunctata* during wet months of the year enhances seed set threefold. During the dry months, the ant *Camponotus planatus* provides a variable and generally less effective defense. Plant

Table 7-6 Quality of ant defense of *Calathea ovandensis* flowers in southern Mexico

Ant Species	Body Size (mm)	Activity/ Inflorescence[a]	Seeds per Inflorescence[b]
Wasmannia auropunctata	1.0	0.67	28.0[a]
Crematogaster sumichrasti	3.5	0.65	19.2[a,b]
Solenopsis geminata	3.5	0.70	16.0[b,c]
Brachymyrmex musculus	1.5	0.66	15.8[b,c]
Monacis bispinosus	5.0	0.51	15.5[b,c]
Paratrechina spp.	2.0	0.58	15.1[b,c]
Pachycondyla unidentata	7.0	0.43	12.1[b,c]
Pheidole gouldi	4.0	0.57	10.3[c]

Source: After Horvitz and Schemske (1984).

[a]Proportion of census dates seen.

[b]Means with the same letters as superscripts are not statistically distinguishable.

protection hinges on the competitive struggle between two ant defenders, both of which are attracted to the same nectaries. An effective defender *(Wasmannia)* wins in wet weather, a less reliable defender *(Camponotus)* is dominant during dry weather.

The reliability of ant defense is often compromised by conflicts of interest. Some ants that defend nectar-producing plants also defend sugar-secreting herbivores. In the mountains of southern Mexico extrafloral nectaries of the perennial herb *Calathea ovadensis* are tended by several species of ants, many of which also tend the specialist herbivore *Eurybia elvina* (Horvitz and Schemske, 1984). *Eurybia* is a moth, the caterpillars of which secrete a fluid rich in sugars and amino acids while they feed on buds, flowers, and fruits. Ants eat this fluid as well as plant nectar, and protect the *Calathea* herbivore along with *Calathea* itself. Horvitz and Schemske found that inflorescences produced few seeds (average 6) when *Eurybia* caterpillars were present and ants were excluded, and many more seeds (average 21) when ants were present but caterpillars were not. Inflorescences with both ants and caterpillars had intermediate (average 14) seed production. Ants protected *Calathea* inflorescences from flies and other seed predators, but they indirectly lowered seed production by 33% by allowing *Eurybia* to be present.

Ant species varied in their effectiveness as protectors of *Calathea ovadensis*. Different species occupied this plant at different sites in the mountains of southern Mexico. Influences of most ant species were indistinguishable from each other (Table 7-6), but did increase seed production over plants without ants at all. Two ironies illustrated how difficult field tests can be. Ant effectiveness at defense has usually been estimated by most ecologists as the intensity of defense against artificial "intruders" (forceps, fingers, twigs). On this score, Horvitz and Schemske found that *Pheidole* was by far the most aggressive ant, although it was clearly least effective against insects. Moreover, a conflict existed between protecting and dispersing seeds. *Wasmannia,* the best promoter of seed *production,* actually depressed seed *dispersal. Calathea* is an ant-dispersed plant with a nutritious aril as an inducement to dispersal. *Wasmannia* chewed off the aril without dispersing the seed.

The answer to the question "Why aren't protective ant mutualisms universal?" is probably "Ants aren't consistently reliable." Especially effective ant protectors may not be present in some places, either by chance or because climates at high latitudes or altitudes exclude them. More to the point, ants may be seduced by destructive herbivores that offer them rewards equal to or better than those provided by the plant. In some cases, an ant that is beneficial to plants at one stage, like *Wasmannia* on *Calathea,* may be detrimental at other stages.

SUMMARY

The evolutionary ecology of nonsymbiotic mutualisms illustrates the reciprocal exploitation of plants and animals. Cost/benefit analyses of foraging behavior spell out the adaptive opportunities and constraints for animals that must select among a variety of flower or fruit resources. A cost/benefit analysis also helps to define mating and dispersal strategies of plants that must make use of visitors that may pollinate their flowers or disperse their seeds, or may destroy flowers or seeds. Protective ant mutualisms illustrate in detail the subtle conflicts of interest of interacting species that must be accommodated in mutualisms. Most nonsymbiotic mutualisms involve animals that interact with several species of plants, and plants that depend on several species of animals. Pairwise relationships are rare, and are usually attributable to intense ecological interactions in species-poor habitats or to other special circumstances.

Mutualisms are widespread in nature because exploited species often evolve means of taking advantage of their exploiters. Mutualisms often grade imperceptibly into antagonistic relationships such as seed predation, parasitism, and herbivory. Far from being anomalous in Darwin's world of universal "struggle for existence," mutualisms illustrate how plants and animals cope with, and often take advantage of, the destructive potential of their biological environments.

STUDY QUESTIONS

1. How would you expect the foraging tactics of butterflies and bees to differ?
2. How might a plant evolve responses to frequent visitation by highly social honeybees as compared with visitation by solitary bees of the same size?
3. What difference does it make to plant fitness if some pollen donors are more effective than others, if all ovules produce seeds?
4. Would you expect fruit-eating animals to specialize on a single species of fruiting plant? Why or why not?

5. Many seeds eaten by monkeys or large terrestrial mammals are defecated in viable condition, but neither show seed dormancy nor extreme vulnerability to seed predators. What is the advantage in dispersal for such seeds?
6. Unaggressive ants use plant domatia, nectar, and sometimes food bodies, but do not defend the plants against herbivores. The cheaters thereby save energy and avoid the risk that protectors expend on plant defense. What keeps cheater species from dominating ant-plants, and thereby destroying ant-plant mutualisms?
7. What keeps ants from being as reliable for plant defense as alternative modes of defense, such as secondary compounds?

SUGGESTED READING

Several anthologies provide stimulating discussions of mutualisms, as well as access to a wider literature. *Coevolution*, edited by Douglas Futuyma and Montgomery Slatkin (1983), and *The Biology of Mutualism: Ecology and Evolution*, edited by Douglas Boucher (1985), provide general coverages of theory and practice in the study of mutualism. Excellent discussions of pollination biology may be found in *Handbook of Experimental Pollination Biology*, edited by C. Eugene Jones and R. John Little (1983), and in *Pollination Biology*, edited by Leslie Real (1983). With *Seed Dispersal*, David Murray (1986) has compiled a unique contemporary anthology on the ecology of the subject. Andrew Beattie (1985) gives a thoughtful account of seed dispersal and protection mutualisms involving ants in *The Evolutionary Ecology of Ant-Plant Mutualisms*.

8

Natural Selection and Mutualism

Do mutualists coevolve? It is easy to visualize the reciprocal responses of two species locked into an inescapable struggle for survival, as often occurs among specialized insect herbivores or pathogens and their host plants. There is little doubt that alleles in wheat that confer resistance to the Hessian fly are selected by that devastating pest, or that mutant alleles in the fly that counter wheat defenses are favored because they permit the flies to reproduce on a preferred host. Pairwise coevolution is more difficult to visualize when selection is intermittent and complicated by interactions with more than the two species of interest. Agents of pollination and seed dispersal rarely have close physical associations with flowering and fruiting plants. Most flower visitors and fruiteaters use several to many plant species, and most plants are visited by several to many animal mutualists. Diffuse coevolution, defined earlier as the response to selective pressures generated by two groups of species on each other, may seem a more appropriate concept for generalized interactions between herbivores and plants (e.g., grazers and grasses) and for nonsymbiotic mutualists.

This chapter explores possible modes of mutualistic evolution. The first step toward an evolutionary analysis of mutualism is to establish the conditions for coevolutionary change among nonsymbionts that are only partially dependent on each other, such as plants and their pollinators. The second is to investigate empirical evidence that such changes occur. The third is to provide an overview of possible scenarios of coevolutionary change among nonsymbionts. This analysis does not have the benefit of a long history of agricultural research, which has been instrumental in understanding the genetics of coevolution among plants and herbivores. However, a recent surge of interest in mutualisms promises important new insights into the processes and patterns of coevolutionary change.

THEORY

At least three conditions are required for coevolution to occur (Table 8-1). (1) A persistent relationship must exist between two species, or two groups of

Table 8-1   Conditions for the coevolution of mutualists

1. Populations consistently interact in ways that benefit each other's survival and/or reproduction.
2. The differences in behavior, physiology, or morphology that mediate mutualistic relationships are heritable.
3. Population structure permits evolution of local mutualisms.

species, that benefit each other in distinctive ways. Relationships between animals and plants that are observed at any one place and time often do not result from a common evolutionary history (Janzen, 1980). Chance associations of compatible species are undoubtedly common in nature. These result from parallel or convergent evolution and ecological coincidence, not coevolution. (2) Variations in the attributes that directly affect the success of mutualism, such as bee foraging behavior or flower number and nectar content, must be influenced by gene differences. Environmental variation will preclude evolution, or coevolution, if the genetic influence on characters affecting mutualism is weak (Appendix I). (3) Fragmented species populations allow genetic drift and local adaptation to maintain alleles beneficial to mutualism when selection is intermittent or weak, as it often is between nonsymbiotic mutualists.

Interactions

One approach to the evolutionary study of mutualisms is to model and then measure the benefits and costs of interactions for the fitnesses of the participants. In general, a relationship should evolve if benefits in fitness exceed the costs for both species (Keeler, 1985). A complete analysis for any given mutualism would require (1) identifying costs and benefits, (2) using a common currency associated with fitness (e.g., energy gain and loss), (3) measuring cost, benefit, and fitness, and (4) solving a set of simultaneous equations of all such variables for interacting species. A complete analysis has never been accomplished; however, Kathleen Keeler has specified theoretical conditions required for mutualism for several ecological relationships. One of her examples illustrates the approach.

Imagine the conditions required to make **myrmecochory** (seed dispersal by ants) beneficial for an herbaceous plant of the forest floor. Cooperating herbs provide food bodies (elaiosomes) on their seeds that induce ants to transport and bury seeds. For myrmecochory to persist, the fitness of individuals that invest in elaiosomes (fitness of the mutualists or w_m) must be greater than that of those that do not (fitness of the nonmutualists or w_{nm}), i.e., $w_m > w_{nm}$.

For those investing in myrmecochory:

$$w_m = pw_t + qw_n$$

where w_t is the fitness of individuals with elaiosomes and transported seeds, and w_n is the fitness of individuals with elaiosomes and nontransported seeds.

Here p is the frequency of plants with successfully transported seeds, and q is the frequency of plants with untransported seeds ($p + q = 1$).

Fitnesses for individuals with elaiosomes and successfully and unsuccessfully transported seeds are defined as

$$w_t = l_t v_t - I_m$$
$$w_n = l_n v_n - I_m$$

where l_t and l_n are the percentages of seeds not lost to predation in each category, v_t and v_n are the percentages of seeds that germinate in each category, and I_m is the investment in elaiosomes that could have been investment in seeds. For unsuccessful myrmecochores, l_n and v_n may be low because of nutrient limitations under the parent, dryness, or predation of seeds by insects or rodents that forage near fruiting herbs.

For myrmecochores with elaiosomes, substitution and simplification allows fitness to be defined as

$$w_m = p(l_t v_t) + q(l_n v_n) - I_m$$

while for nonmyrmecochores without elaiosomes,

$$w_{nm} = l_n v_n$$

If $w_m > w_{nm}$, then

$$p(l_t v_t) + q(l_n v_n) - I_m > l_n v_n$$

which reduces to

$$p(l_t v_t) - I_m > p(l_n v_n)$$

According to Keeler's model, for mutualism to occur (1) benefit must cover cost of the food body, (2) $v_t > v_n$, and/or (3) $l_t > l_n$. In Keeler's model, the p's on either side of the inequality are small and do not influence the outcome. A parallel set of equations could be constructed for each interacting ant species.

The strength of a cost/benefit analysis is that correlates of fitness, such as seed production, seed germination, and seedling survival can often be measured in nature. An additional strength at the present stage of knowledge is that such analyses do not require specific genetic models. A qualitative assessment of the strength or weakness of an ant/plant mutualism can be made without knowing allele frequencies, dominance within loci, interactions between loci, or population sizes. Genetic and population variables are difficult to measure in nature. Weaknesses of cost/benefit analyses are that indirect measures of fitness are inexact. Moreover, evolution may be promoted or constrained by just those genetic and population variables that cost/benefit analyses omit. Precise quantitative predictions require genetic information.

Inheritance and Mutualism

Natural selection is a potent evolutionary force only when gene differences underlie the phenotypic variation of traits. Rather little is known of the genetics

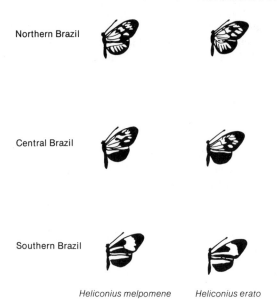

Northern Brazil

Central Brazil

Southern Brazil

Heliconius melpomene   *Heliconius erato*

Fig. 8-1 Müllerian mimics in wing coloration of the butterflies *Heliconius melpomene* and *Heliconius erato*. Shown are strongly convergent morphs of each species in three localities in Brazil. Mimicry complexes offer well-studied examples of coevolution of traits with simple Mendelian inheritance. Because both of these species are distasteful to predators, the warning coloration of each species reinforces avoidance of similar wing patterns in the other species. After Turner (1981).

of flowering and fruiting in wild plants, and virtually nothing is known of the genetics of foraging and feeding by pollinators and dispersal agents. Genetic mechanisms are important because **genetic architecture,** or dominance within loci and interactions between loci (Appendix I), influences the rates of coevolutionary change under the weak or intermittent selection likely to be common in nonsymbiotic mutualisms (S. A. Levin, 1983). An example of a classic mutualism between animal species, Müllerian mimicry, illustrates how dominance in simple Mendelian traits influences coevolution.

Alan Templeton and Lawrence Gilbert (1985) consider two tropical *Heliconius* butterflies that mimic each other in coloration, and compete for food from the same passion flower *(Passiflora)* flowers. The butterflies are both distasteful to predators, and both possess simple Mendelian loci that determine patterns of bright **aposematic** wing coloration that warn would-be predators of insect toxicity (Fig. 8-1). Wing traits coevolve because predators that taste one species learn to avoid both. Because the genetic architecture of *Heliconius* wing patterns is known, Templeton and Gilbert can construct plausible genetic selection models of coevolution. Models of coevolution of Mendelian traits in two animal species offer insights into as yet uninvestigated genetic coevolution of animal and plant mutualists.

For wing traits, Templeton and Gilbert propose alleles *A* and *B* that confer resemblance, with recessives conferring detectable but lesser resemblance, in two species of *Heliconius* butterflies. The fitness of aa individuals is

$$ag_{aa}N_1 + b(1 - g_{aa})N_1 + c(1 - g_{bb})N_2$$

where g_{aa} is the genotype frequency of aa in species 1, g_{bb} is the genotype frequency of bb in species 2, N_1 and N_2 are the densities of species 1 and 2, and a, b, and c are constants that measure the proportional fitness of aa individuals from individuals of the same phenotype, the other intraspecific phenotype, and the dominant phenotype of the other species 2, respectively. Resemblance functions are $a > b > c$. The fitness of the dominant phenotype A (either AA or Aa) in species 1 is:

$$a(1 - g_{aa})N_1 + bg_{aa}N_1 + s(1 - g_{bb})N_2 + dg_{bb}N_2$$

where s and d measure the proportional fitness contributions to A individuals from a resemblance to the dominant and recessive phenotypes in the other species, respectively, with resemblance assumptions $a > s > b$ or d. Similar fitness equations can be written for species 2.

The simplest assumption for evolution of wing traits is that the N of each species is unaffected by either trait. The further simplification that N is constant for each permits a quick calculation of gene frequency change.

If p is the frequency of the A allele, the change in allele frequency over one generation *(P)* is

$$\Delta P = pa_A/w$$

where w is the average fitness of the population (see Appendix I) and a_A is the average fitness excess of A, which is the average mean fitness of individuals bearing A minus w. Because p (frequency of the A allele) and w (population fitness) are always positive, the direction of evolution is determined by the sign of a_A. To calculate a_A,

$$g_{aa}[N_1(1 - 2g_{aa})(a - b) + dg_{bb}N_2 + (1 - g_{bb})N_2(s - c)]$$

A will always increase in frequency when $g_{aa} < 1/2$. Templeton and Gilbert point out that the increase in the A allele occurs because it is the dominant warning coloration. Even if no interspecific interaction between butterflies occurs ($N_2 = 0$ or predation does not occur), p will increase when $g_{aa} < 1/2$. When $g_{aa} \geq 1/2$, p increases only with very strong interspecific interactions. Coevolution will occur because of mimicry only when p is small and the advantage is large.

This Mendelian model specifies conditions under which coevolution can occur. Templeton and Gilbert suggest other scenarios that assume different effects of predation or competition on population densities, which were kept constant above. If the genetic basis of relevant traits, allele frequencies, selection coefficients, and population sizes are known, one can make quantitative predictions about the rate and direction of coevolution and test them in nature. This is the ultimate objective of evolutionary models.

Theoretical models with this degree of realism are not presently possible for nonsymbiotic mutualisms between animals and plants. One reason is that the genetics of relevant traits is rarely known (Schemske, 1983). A second reason is that the selective influence of one species on another is often unpredictable in animal and plant mutualisms. Predation maintains *Heliconius* mimicry in an unusually predictable way. Imagine a less certain interaction between two non-symbiotic mutualists, a pollinator species and a flower. Two loci determine whether columbine flowers have long spurs and are therefore accessible only to long-tongued hawkmoths, or are spurless and therefore accessible to any hawkmoths (Prazmo, 1961). A mutant family of spurred columbine flowers in a patch of spurless individuals is far more likely to attract long-tongued hawk-moths than force the evolution of long tongues among short-tongued insects. A mutation with a major effect on a Mendelian trait such as spur length is more likely to disarticulate an existing mutualistic relationship and promote a new one than promote coevolution. In this example the selective influence of short-tongued moths on columbines abruptly disappears, and the selective influence of long-tongued moths just as abruptly increases in force. Selective scenarios for mutualisms involving more complex genetic architectures are virtually unexplored.

How might diffuse coevolution occur? **Guilds** are groups of structurally and functionally similar animals or plants. Often a pollinator guild (e.g., medium-sized bees) includes species with approximately equivalent effects on the trans-fer of pollen in a plant guild (e.g., asters), which in turn provides roughly equivalent nectar and pollen resources. Plants may adapt to a guild of pollina-tors or dispersers rather than to particular species, and the animals in turn adapt to a guild of plants with similar resources, rather than to characteristics of any single plant species. What will evolve is a generalist tactic of using a guild of similar animals, or similar plants, with equivalent effects on fitness (Levins, 1968). The degree of generalization depends on the taxonomic breadth of mu-tualists that have more or less equivalent effects on the fitness of the species with which they interact. For some plants, any hummingbirds may be suitable pollinators. For others, only long-billed species will do. Genetic models of traits evolved through diffuse coevolution will differ little from the general selection model (Appendix I), although the source of selection may be broad (e.g., insects and hummingbirds) or narrow (e.g., four species of long-billed hummingbirds). Clearly the appropriateness of the term coevolution diminishes as the various sources of biotic selection broaden.

Population Structure and Mutualism

Population structure refers to the extent to which large populations are di-vided into smaller subunits. Species populations may be genetically continuous, or may be subdivided into large or small local subunits (demes). Population structure is important because genetic evolution involving rare or recessive al-leles may be much faster among many small demes than in a large continuous population (Wright, 1980). This occurs for two reasons.

First, alleles become common or fixed by the chance effects or inbreeding of close relatives much more easily in small (2–100) than in large (>100) populations. Random, nonselective change in gene frequences, called genetic drift, is likely to be especially important among self-compatible plants or plants with limited pollen dispersal. The rate of inbreeding *(F)* is

$$F = 1/2N_e$$

where N_e, termed the effective breeding population, is the number of individuals that actually reproduces. Genetic drift in small demes of structured populations maintains generally rare or recessive alleles at levels higher than would be likely if they were constantly masked by the influx of alleles from other parts of a species range.

Second, the strength of natural selection often varies from one deme to another. If genetic drift has produced a wide variety of distinctive gene combinations in small demes, selection has the opportunity to act on locally common alleles that are recessive or rare in the total species population (Templeton and Gilbert, 1985). In large, genetically continuous populations (large N_e) genetic drift is negligible and gene frequency changes due to local selection are overwhelmed by immigration from elsewhere. When N_e is large, gene flow between sites swamps genetic differences among sites.

Plant ecologists often find that gene flow between local plant populations is restricted. Under any but the most unusual circumstances (e.g., seed dispersal by water), most pollen grains and seeds fall close to the parent plant, while a minority are dispersed substantial distances. Particular patterns of animal-mediated gene flow in plants depend on (1) flight patterns of pollinators, (2) rates of pollen deposition, (3) pollen success in fertilization, (4) patterns of seed deposition, and (5) seedling success.

Gene flow through pollen movement is often limited. Donald Levin and Harold Kerster (1974) review a number of studies showing that bees often forage within the same 10 to 20 m^2 patch of flowers day after day. The dispersal of pollen depends on pollinator flight patterns and on the proportion of pollen grains left on each successive flower visited (pollen carryover). Pollen viscosity, which varies from plant species to species, and pollinator morphology and behavior determine whether a large or small proportion of the pollen rubs off onto stigmas at each visit. When a constant proportion of pollen is deposited at each flower visit—a rate of deposition termed exponential decay (Fig. 8-2)—most or all of a pollen load may be left on a single individual plant with many flowers, or on a nearby neighbor. If a smaller proportion is dislodged on each visit, pollen dispersal distance and consequently outcrossing increase. Increasing distance of pollen dissemination also increases the danger that the pollinator will visit inappropriate species or eat the pollen.

Levin and Kerster (1968) determined the deme structure of a common American wildflower, *Phlox pilosa*. Assuming exponential decay of pollen loads, they demonstrated that butterflies rarely carried phlox pollen more than 3 m from a patch. Rare insect flights permitted potential pollen dispersal to 15 m (Fig. 8-3). Seed dispersal in this species is even more restricted because the

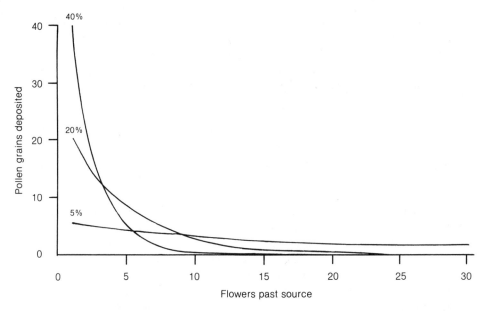

Fig. 8-2 A computer model of pollen dispersion when 40, 20, or 5% of the 100 pollen grains carried by a pollinator is left on each successive flower visited. This simulation suggests that virtually all pollen that is easily rubbed off is left close to the source flower. In plants with many flowers, much of the pollen will be left on the source plant itself. The less likely pollen is to be left on a plant (i.e., the more viscous the pollen), the more likely it is that it will be deposited at some distance from the source flower and, consequently, the source plant. Eventually the advantages of outcrossing are countered by an increasing likelihood that pollen carried through many successive visits will be deposited on plants of other species, or eaten by insect or bat pollinators. After Lertzman and Gass (1983).

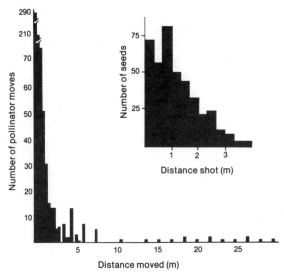

Fig. 8-3 Potential gene dispersal in *Phlox pilosa*. Most butterfly flights are less than 3 m from a phlox plant, although some are much farther. Inset: Seed shadow of *Phlox pilosa*. After Levin and Kerster (1968).

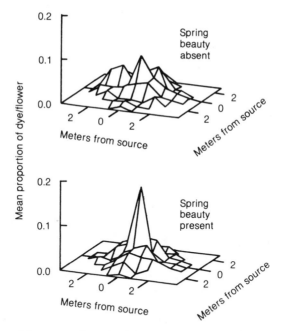

Fig. 8-4 Patterns of fluorescent dye carried by pollinators from potted chickweed *(Stellaria pubera)* plants. The dye simulates pollen dispersal. The upper graph shows dye movement away from plants in single-species stands. The lower graph shows much more restricted dye movement in mixed stands of chickweed and spring beauty *(Claytonia virginica)*, which is a competitor for pollinators. Each graph depicts mean distribution of dye particles in four replicate plots containing a total of 24 source plants. After Campbell (1985).

seeds are explosively discharged only a few centimeters from the fruit capsule. As a result, phlox populations are split up into small effective breeding populations, called **neighborhoods.** Wright (1946) defines the neighborhood area for a hermaphroditic plant population as N_e/d, where d is the density of flowering plants $N_e = 12.6s^2d$ when both male and female gametes (pollen and seeds) have equivalent dissemination, and s^2 is the variance of the pollen and seed distributions. With modifications for differences in pollen and seed distributions, Levin and Kerster calculate that N_e for *Phlox pilosa* is 75 to 282 individuals in a neighborhood area of 11 to 21 m^2. Barring rare insect odysseys, each patch of these and other small herbs is a genetic island unto itself.

Biotic factors influence neighborhood sizes for different populations of the same plant species, and for plants with widely divergent reproductive biologies. For instance, chickweed *(Stellaria pubera)* competes with spring beauty *(Claytonia virginica)* for bee fly and bee pollinators. Campbell and Motten (1985) found that the neighborhood size (including 95% of pollinator moves) for chickweed was 16 individuals when spring beauties were absent, but was reduced to 12 when spring beauties were present to compete for bee fly pollinators (Fig. 8-4). Competition significantly altered chickweed population structure.

Populations of large plants may be less subdivided than those of herbs like phlox and chickweed (Levin and Kerster, 1974). Bats, birds, and large insects that pollinate many trees are strong fliers that visit widely separated plants. More importantly, seeds of large shrubs and trees are often disseminated by birds and mammals that carry seeds dozens to thousands of meters. Long-distance pollinators and wide seed dissemination lead to gene exchange among widely separated individuals, tending to overwhelm genetic evolution due to weak local selection or genetic drift.

PATTERNS IN NATURE

The study of the evolution of mutualisms in nature is a study of paradoxes. Patterns of morphological divergence and convergence are common among flowers, fruits, and vegetative traits that promote ant mutualisms, just as they are common among animals that visit or tend plants. On the other hand, the evolution of plants and animals may occur at different rates. Fossil evidence indicates remarkable stability of some plant taxa over millions of years, but far greater change among animal mutualists during the same periods. What do patterns of morphological divergence, convergence, and stability suggest about the evolution of mutualisms?

Divergence and Convergence

One of the most striking features of mutualisms is the morphological malleability of flower and fruit characteristics used to attract pollinators and dispersal agents. Unrelated plants often converge in the same pollination and dispersal syndromes (see Tables 6-2 and 6-4), while some taxonomic groups have members in each of many pollination and dispersal syndromes. Traits that influence pollination or seed dispersal are evidently malleable over evolutionary time.

Verne and Karen Grant (1965) have provided a classic study of pollination biology in the phlox family (Polemoniaceae). With 327 species and 18 genera, the phlox family includes an astonishing array of modifications for pollination by widely differing animals (Fig. 8-5). The genus *Polemonium* is close to the ancestral bee-pollinated form. Other genera include species adapted for pollination by butterflies (many *Phlox* species), mountain-living noctuid moths (*Phlox caespitosa* and others), hummingbirds (some *Cantua, Gilia, Ipomopsis,* and others), hawkmoths *(Ipomopsis),* bee flies with extremely long proboscises *(Linanthus),* beetles *(Ipomopsis congesta* and *Linanthus parryae),* and even bats *(Cobea).* This list includes several genera with species adapted to very different pollinators (e.g., *Ipomopsis* and *Linanthus*).

Fruit sizes, shapes, and nutritional rewards also show tremendous potential for evolutionary modification. A large and cosmopolitan group like the bean family (Leguminosae; 17,000 species) includes fruits and seeds adapted for

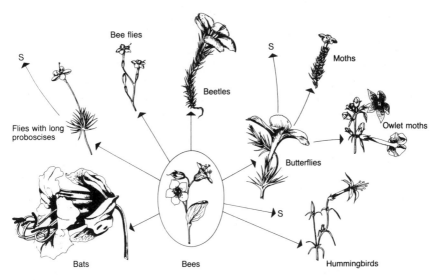

Fig. 8-5 The phlox family (Polemoniaceae) illustrates divergence of flower form due to selection from different pollinators. Self-pollination (S) evolved independently in several of these adaptive branches. From Barth (1985); after Grant and Grant (1965).

dispersal by gravity, wind, and water, by fruit-eating birds, bats, and large mammals, by nut-hoarding rodents, and passively by fur or feathers (van der Pijl, 1972). Even a small family, such as the tropical Lecythidaceae (450 species), includes species that have modified the basic structural plan of a woody fruit to attract birds, primates, fish, rodents, and even perhaps large ungulates, or to be disseminated by wind or water (Prance and Mori, 1978). It is not yet known whether fruits show local adaptations to visitor assemblages.

Studies of convergence and divergence in the traits that pollinators and seed dispersers use to feed are not as comprehensive as botanical analyses of flower and fruit form. Bill length in hummingbirds (Stiles, 1975) and tongue length in nectar-feeding insects (Heinrich, 1979b) both show tremendous divergence in closely related animals. On the other hand, a convergence of long hummingbird bills and long hawkmoth tongues allows these two unrelated animals to sip nectar from the same flowers. More comprehensive analyses of divergence and covergence in morphology, foraging behavior, and digestive physiology offer an open field for future investigations of the evolution of mutualism.

Time Scales

Incomplete though it is, the fossil record offers insights into the tempo of evolution among plants and potential bird or mammal dispersal agents. Many plant families and genera that had evolved by the early Tertiary (60 million years ago) are still alive today. Fossils of known ages can calibrate rates of chromosomal evolution of living species, thereby giving an estimate of longevity of fossil species not alive today (Levin and Wilson, 1976). Stebbins (1982), for instance, found that 54 California plant species are known to have survived, with negligible modification, for 7 to 55 million years. In a broader analysis, Carlos Herrera (1985) surveyed both fossil and genetic evidence to find that mean species durations of angiosperm shrubs and trees are 27 and 38 million years, respectively. In contrast, species durations averaged 0.5 to 4 million years for mammals and 0.5 million years for birds. Plants apparently generate new species rapidly through changes in chromosome number and arrangement. Once established, a plant species often lasts a very long time. Mammals and birds undergo adaptive radiation, but as established species they come and go much more quickly than shrubs and trees.

The historical record is coarse, and errors may exist in comparisons of the species longevity of plants and animals. Paleobotanists recognize plant species largely on the basis of fossil fruits, while paleozoologists use characters unrelated to fruit-eating. Animal traits relevant to frugivory may be as stable as those of fruits themselves, but less stable than characters used by paleozoologists. Nonetheless, the stability of fruit structure over millions of years probably indicates that many shrub and tree species have outlived successive assemblages of dispersal agents. David Snow (1981) found that many African trees evolved long before any of their contemporary dispersal agents existed. Paul Cox (1983) noted that museum mounts of extinct Hawaiian birds (e.g., *Loxoides kona*) are often dusted with pollen of the native ieie vine *(Freycinetia arborea).* Ieie flowers are now pollinated by introduced Indian white-eyes *(Zosterops japonica).* The proportion of plants that survive successive extinctions of pollinators or dispersal agents is not known.

Ecological Dynamics

Intensive ecological investigations of nonsymbiotic mutualisms rarely suggest strong, persistent selection between species that would result in pairwise coevolution. The potential for specialization often exists, but variation in selective histories or strong asymmetries in the value of a relationship to different mutualists preclude stepwise, reciprocating selection like that seen in gene-for-gene coevolution of insects and host plants (Table 8-2). Some examples illustrate the constraints that ecology places on coevolution.

Relationships between nonsymbiotic mutualists vary in space and time. Recall that *Calathea ovadensis,* a large tropical herb discussed in the previous chapter, receives variable protection from ants in the mountains of southern Mexico. This species also provides an excellent example of variability in pollination dynamics. The flowers have a trip mechanism, by which pollen is deposited on visiting insects. Certain bee pollinators *(Bombus* and *Rhathymus)* trip 100% of the flowers of *Calathea ovadensis* that they visit (Schemske and Horvitz, 1984). However, these especially effective bees vary in abundance from one patch of flowers to the next, and from season to season. Most *Cala-*

Table 8-2 Constraints on coevolution between species of plants and their agents of pollination and seed dispersal

1. Pollinators and dispersal agents vary in abundance and effectiveness in space and time.
2. Flowers and fruits vary in abundance and quality in space and time.
3. Long generation times lead to slower rates of plant than animal evolution.
4. Dependence of plants on animals and animals on plants is often asymmetric.
5. Plant and animal traits may differ in their capacity to respond to selection from a mutualist.

thea pollination is actually accomplished by relatively inefficient bees *(Euglossa, Eulaema, Exaerete)* that trip few flowers per visit, but happen to be very common and consequently visit often. Whether or not *Calathea* once coevolved with Mexican *Bombus* or *Rhathymus,* it persists because abundant, if clumsy, pollinators permit it to reproduce when the two most effective bees are rare or absent.

Dispersal agents likewise vary in abundance in time and space. For instance, a tityra that is critical for dispersal of the Costa Rican tree *Casearia corymbosa* in rainforest (Chapter 10) is irrelevant to the dispersal of its fruits in Costa Rican dry forest, where smaller vireos do the job (Howe and Vande Kerckhove, 1979). Similar reliance of shrubs and trees on guilds of frugivores of roughly equivalent efficiency in seed dissemination also occurs in temperate forests and scrublands for plants that depend on small birds (Herrera, 1984b). In both of these examples guilds of small birds interact with guilds of trees and shrubs that bear small fruits.

Ecological variability is of course sharply curtailed for *symbiotic* mutualists that live inside of hosts. Small algae, bacteria, or fungi that inhabit larger hosts often show far less diversification than the hosts themselves (Table 8-3). Evidently inhabitants face far more uniform selective environments than hosts that face the outside world. Does this imply that coevolution among *nonsymbiotic* mutualists is usually adaptation to compatible guilds (not to particular species) of plants, pollinators, or dispersal agents?

Table 8-3 Number of genera and species of inhabitants and of hosts of selected symbiotic mutualisms

Symbioses	Inhabitants		Hosts	
	Genera	Species	Genera	Species
Rhizobium and plants	1	4	600	17,500
Cyanobacteria and plants	2	?	70	150
Cyanobacteria and fungi	9	?	300	13,500
Chlorella and invertebrates	1	1	14	14
Dinoflagellates and invertebrates	2	4	200	650
Ericoid mycorrhizas	2	2	25	2,000
Vesicular–arbuscular mycorrhizas	4	78	11,000	225,000

Source: Original data from many sources compiled by Law (1985).

EVOLUTION OF MUTUALISMS: ALTERNATIVE SCENARIOS

Comparative, historical, and ecological evidence suggests that more than one evolutionary scenario might explain the origin and evolution of mutualisms. Theory suggests that particular plant and animal species may coevolve, or that guilds of functionally similar pollinators or dispersal agents may slowly coevolve with guilds of functionally similar flowering or fruiting plants. Documenting which of these alternative scenarios accounts for the majority of the thousands of mutualistic associations known in nature is one of the exciting challenges of contemporary coevolutionary ecology.

Species Pairs in Species-poor Communities

Coevolution of pairs of species is theoretically possible, but is unlikely in species-rich communities in which many animals select flower and fruit traits, and many plants influence the fitnesses of animal foragers (Howe, 1984a). Might pairwise coevolution occur in depauperate communities, where particular species of plants and animals are forced into ecological associations with little interference from other species? How might one test this **forced association hypothesis,** which implies that most coevolutionary change occurs in isolated, species-poor habitats?

Obligate relationships between nonsymbiotic animals and plants in desert, mountain, or island habitats suggest that pairwise coevolution may occur where low species diversity promotes strong interactions among species. In the example of *Gilia splendens* (Polemoniaceae) mentioned above, at least the plant side of the plant/pollinator relationship shows clear adaptation to local conditions in species-poor environments (Fig. 8-6). A widespread subspecies has a funnel-like corolla 5–6 mm long. It is pollinated by at least one common bee fly, *Bombylius lancifer,* and probably by other bee flies with proboscises of moderate length. In the San Gabriel Mountains of California, however, the long (15 mm) slender corolla of *Gilia splendens* perfectly matches the extremely long proboscis of a locally common fly, *Eulonchus smaragdinus,* which appears to be a regular visitor of this species at this site. Only a short distance away in the San Bernardino Mountains, the Grants (1965) found that *Gilia splendens* had an even longer (20 mm) brilliant pink trumpet-shaped corolla that attracted three species of hummingbirds with bills that matched corolla length. *Gilia splendens* adapts flower structure to meet local selective challenges by pollinators. There is no evidence yet that animals adapt to different corolla lengths. A coevolutionary response to *Gilia* flowers would not be expected in migratory hummingbirds, which forage on many flowers in both their breeding ranges and in tropical wintering sites. A careful study of proboscis length and behavior in the San Gabriel Mountains and in other races of the fly *Eulonchus smaragdinus* might show adjustment to different *Gilia* corolla forms.

A second intriguing example suggests coevolution of both a plant and its

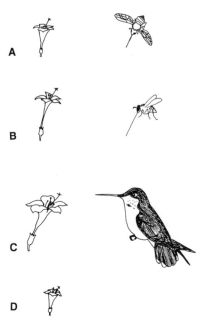

Fig. 8-6 Races of *Gilia splendens* and their pollinators (⅓ life size). (A) Widespread race and *Bombylius lancifer* (Bombyliidae); (B) high San Gabriel Mountain race and *Eulonchus smaragdinus* (Cyrtidae); (C) high San Bernardino Mountain race and *Stellula calliope* (Trochilidae); (D) desert race, largely self-pollinated. After Grant and Grant (1965).

dispersal agent on a remote oceanic island. Stanley Temple (1977, 1979) describes a possible mutualism between the extinct dodo *(Raphus cucullatus)*, a giant flightless bird, and the tambalocoque tree (*Calvaria major*, Sapotaceae) of the biologically depauperate island of Mauritius in the Indian Ocean. The dodo, hunted to extinction by explorers by 1675, was an omnivore that crushed fruits and seeds in a powerful gizzard. Tambalocoque fruits have a fleshy mesocarp and an extremely tough pit that no modern indigenous frugivore can digest. At present, tambalocoques are rare, apparently restricted to ancient trees thought to exceed 300 years in age, and to younger stands propagated by the Forestry Service of Mauritius. The seeds rarely germinate unless they are heavily abraded by Forest Service personnel. Stanley hypothesizes that tambalocoque fruits evolved a nearly indestructible pit in response to seed predation by dodos, and that dodos evolved an unusually robust stone-filled gizzard as a means of eating these once-common fruits. Intriguingly, introduced turkeys crush and digest some seeds, and pass others that germinate shortly after they are defecated.

The simplified selective environments of species-poor communities may permit pairwise coevolution. Such coevolution could be of general importance if associations evolved in geographic isolation later mingle with more complex plant and animal communities. For instance, imagine that long isolation leads to genetic incompatibility between the San Bernardino *Gilia splendens* race and others with which it can now interbreed. If future climatic changes allowed this new *Gilia* species to extend its range to lowland areas that are presently arid, it is conceivable that it could become a widespread species pollinated by many species of hummingbirds with bills of the proper dimensions. This kind of scenario, involving coevolution in isolated species-poor environments followed by range extensions, might account for the *origin* of many plant and animal attributes that are later identified with diffuse relationships.

Coevolution of Guilds

Mimicry complexes demonstrate coevolution of species that share, and through parallel adaptations alter, a common selective environment (Wickler, 1968). Mimicry complexes in related species of distasteful butterflies coevolve wing patterns over broad geographical areas, suggesting that local habitats favor different models and mimics (Benson, 1982). Might guilds of pollinators or dispersal agents and guilds of the plants that they visit slowly adjust to each other, much as distasteful *Heliconius* butterflies evolve warning coloration under selection from predators? Might this **guild coevolution hypothesis** not predict simultaneous and reciprocal evolutionary adjustments of flower or fruit guilds with animals that use them?

At first glance, the fossil record suggests guild coevolution. For instance, between 20 and 5 million years ago guilds of herbivorous mammals evolved grazing dentition in response to the evolution and spread of tough silica-containing grasses. Similarly, changes in fossil flower and fruit morphology over millions of years might imply generalized adjustment to evolving animal assemblages, and vice versa. Unfortunately, the fossil record is too incomplete to show simultaneous reciprocal evolution. Novel flower or fruit forms might have evolved in association with one or a few species of insects or vertebrates in species-poor communities, and later achieved widespread prominence by virtue of preadaptation to similar animal assemblages elsewhere. Usually the alternative scenario of ecological replacement of plant or animal assemblages, without evolutionary change, cannot be excluded. The stability of tree and shrub species over tens of millions of years, as compared with a rapid succession of fossil bird and mammal species (Herrera, 1985), supports the idea that ecological replacement occurs. This **ecological replacement hypothesis,** of course, offers a possible explanation of present associations, not the origins of traits that make those associations possible.

The *co* in coevolution is difficult to document. What one needs to show is simultaneous and reciprocal evolutionary change in groups of plant species and the assemblages of animals that visit them. For instance, the well-documented adaptations of flower morphology in the phlox family to prevailing conditions of pollinator availability (Grant and Grant, 1965) would have to be matched with documentation of novel adaptations among pollinator guilds in behavior, proboscis length, or sensory adaptations to the new flower forms. This is, once again, a challenge for the future.

SUMMARY

Coevolution of nonsymbiotic mutualists is more difficult to visualize than that of pathogens or specialized insects and their hosts because selection among mutualists often appears to be weak, intermittent, and confounded by several interacting species. Modes of mutualistic coevolution are not well understood. Theoretical approaches suggest what is possible; preliminary empirical studies suggest what is probable.

Models of coevolution emphasize phenotypic interactions between species or explicit genetic models. Cost/benefit analyses define ecological advantages and disadvantages of mutualisms as judged by indirect measures of fitness, such as reproductive success or energy gain. Explicit genetic models may use simple Mendelian or quantitative traits, and may explore interactions between two species or guilds of functionally similar plants or animals. Most models are designed to explore the evolutionary effects of two species on each other, assuming that one or two loci influence the traits of most direct importance to the mutualism. These are valuable in understanding mimicry, and will clarify animal and plant coevolution involving Mendelian traits. Promising approaches incorporate polygenic inheritance, population structure, and the effects of genetic evolution on such ecological variables as population size. Genetic models suggest that both pairwise coevolution and diffuse coevolution of species guilds are possible, but the relative importance of each is not yet known.

Alternative scenarios suggest different modes of coevolutionary change among nonsymbiotic mutualists. Coevolution of pairs of plant and animal species is most likely in species-poor communities where interactions are not confounded by high species diversities of pollinators, dispersal agents, and plants on which they depend. In such cases, models of interactions between pathogens or insects and their hosts plants have some relevance. Coevolution of functionally similar guilds of animals with functionally similar guilds of plants is more likely in species-rich communities. The most appropriate analogue in herbivory involves relationships between animals that feed on many plants (e.g., grazers) and the dozens or hundreds of plant species that they eat (e.g., grasses). At present it is impossible to know whether pairwise coevolution in isolation, followed by range extentions into species-rich communities, or reciprocal evolution of guilds best accounts for the origins of the thousands of nonsymbiotic mutualisms that surround us.

STUDY QUESTIONS

1. How does animal mobility influence coevolution of nonsymbiotic mutualisms?
2. What is meant by an asymmetrical mutualism? How do ecological asymmetries influence coevolution?

3. Algae and fungi that live within other organisms rarely diversify, while the fungi, trees, or invertebrates that they inhabit often include hundreds or even thousands of species (Table 8-3). Suggest at least two hypotheses that might lead to an explanation of this disparity.

4. Models of mimicry are often used to suggest how nonsymbiotic mutualisms might evolve. Why might mimicry be an inappropriate model for plant and animal coevolution?

5. Pollination biologists have a long tradition of research in plant population structure, while ecologists interested in herbivory do not. Why might this be so? How does selection by herbivores differ from selection by pollinators?

6. Devise a hypothetical cost/benefit model for an interaction between *Adrena rozeni* and the Mojave desert primrose, *Camassonia claeviformis,* discussed in the previous chapter.

7. Distinguish between the forced association, guild coevolution, and ecological replacement hypotheses for the origin of mutualisms.

SUGGESTED READING

Douglas Futuyma and Montgomery Slatkin (1983) edit a series of critical appraisals of coevolution in their anthology *Coevolution.* Contributors to the volume *The Biology of Mutualism: Ecology and Evolution,* edited by Douglas Boucher (1985), offer numerous important insights into the biology of mutualisms.

IV

ANCIENT AND
MODERN COMMUNITIES

An ecological community is an assemblage of species living in the same place at the same time. Community ecology is the study of processes that influence the distribution and abundance of species in different communities. Patterns of immigration, extinction, biological interaction, and the interaction of species with the physical environment all contribute to community identity and community change.

Plant and animal relationships play decisive roles in shaping ecological communities. Herbivores may determine which plant species succeed or fail in a community, either by exterminating some species directly or by shifting the competitive balance in favor of some species over others. The presence or absence of pollinators, dispersal agents, or ant mutualists likewise determines whether some herb, shrub, and tree species survive and reproduce effectively in any given place and time. In turn, vegetation makes animal life possible. The success or failure of browsing mammals, sapsucking insects, or fruit-eating bats depends on the availability of leaves, stems, flowers, or fruits suitable for food.

Part IV explores the ways in which plant and animal relationships influence community composition. Broad outlines of historical relationships are well known, and recent refinements in the analysis of fossil communities help us understand modern assemblages of animals and plants in light of evolutionary history. Moreover, a rapidly expanding ecological literature on herbivory and mutualism gives modern ecologists unprecedented opportunities to explore and preserve contemporary communities.

9
Ancient Communities

Imagine the entirely green world of 145 million years ago (Fig. 9-1). No flowers brighten the landscape, no red or blue fruits dangle in the underbrush. The
forest is a tangle of rasping tree ferns and gymnosperms. Butterflies do not sip
nectar from deep corollas. Neither butterflies nor corollas exist. Cockroaches
quietly carve up and chew bits of fronds. No birds or bats hop or flutter in the
treetops, nor do monkeys chatter from the forest edge. Herbivorous reptiles
solemnly chew the foliage, keeping a wary eye out for reptilian carnivores.

Fig. 9-1 A late Jurassic scene, 145 million years ago. Two *Stegosaurus* are shown foraging
for plants in a primeval forest. One of many genera of herbivorous dinosaurs, *Stegosaurus* is
distinctive for the armored plates that protect its vertebral column from predators. From a
mural by Charles Knight, Courtesy of Chicago Field Museum of Natural History.

Table 9-1   An outline of the fossil record

Era	Beginning of Era or Period (millions of years ago)	Period	Events
Cenozoic	1.7–2.5	Pleistocene	Ice ages; evolution of humans
	70	Tertiary	Diversification of modern birds, mammals, insects, and flowering plants
Mesozoic	135	Cretaceous	Diversification of flowering plants; origin of some modern orders of mammals; most modern orders of insects present; abrupt extinction of dinosaurs and many insect groups
	190	Jurassic	Peak of dinosaurs; appearance of modern conifers; origin of birds, mammals, flowering plants, and most modern insect orders; extinction of many insect groups
	225	Triassic	Dominance of mammal-like reptiles; diversification of insects
Paleozoic	270	Permian	Appearance of modern insect orders; extinction of primitive insects and early amphibians
	350	Carboniferous	Vast coal-forming fern, seed fern, and horsetail forests; origin of some modern insect orders, including flying forms; diversification of fossil insects, including giant forms; dominance of amphibians; origin of reptiles
	400	Devonian	Earliest seed plants; origin of wingless insects and amphibians; dominance of fishes
	440	Silurian	Earliest land plants; earliest land animals (millipedes); diversification of fish
	500	Ordovician	Origin of vertebrates; diversification of marine invertebrates
	600	Cambrian	Appearance and diversification of invertebrates; first arthropods (e.g., trilobites)
Precambrian	3,500		Origin of life; microscopic fossil cells; origin of invertebrates

This is a world in slow motion, all the slower on chilly nights when its cold-blooded inhabitants drift into a dull torpor.

Alien as these images seem, they represent far more history than the few million years that humans or their apelike ancestors have walked the Earth (Table 9-1). In fact, even the dominance of reptiles, 270 to 100 million years ago, is a blink of the eye in the history of the Earth. Imagine that the 4.5 billion year history of this planet were compressed into the scale of a single year. Recognizable invertebrates only appeared on October 15, the reptiles reigned from mid to late December, the angiosperm revolution occurred on Christmas Day, and *Homo sapiens* appeared at 10:50 PM on December 31. All of written human history occupies only the last half minute of this fanciful year. Of concern here is the frenzied evolutionary activity of this cosmic holiday season.

This chapter has dual objectives. First, it reviews key interactions between plants and animals in the broad sweep of the history of life. This analysis must be general, because early fossil records are too coarse for detailed ecological analysis. Second, this chapter draws on the much more complete fossil records of the last 20 million years to explore specific historical hypotheses. In particular, it explores the ecological explanations for the pulses of extinction of vast assemblages of large mammals in North America. These culminated in the Pleistocene disaster 10,000 years ago that stripped North America of all but a few of the huge mammals, the mammalian **megafauna,** that once roamed the entire Western Hemisphere. We then explore the hypothesis that the megafauna shaped the evolution of many plants that are still alive, leaving anomalous botanical anachronisms reflecting conditions that no longer exist.

ANCIENT HISTORY

Fossilized hard parts of ancient organisms give us a tantalizing record of the history of life. Unusual circumstances make this record possible. Animals or plants with hard tissues must die or settle in fine sediments where they are protected from the ravages of decay, scavengers, and the destructive forces of wind and water. Lignified plant tissues or bone must then be permeated with mineral-bearing water, forming stony structures that we call fossils. The fossil deposits must by chance avoid destruction over millions of years of erosion, mountain building, and other natural forces. Some deposits must be located near the surface of the Earth, where paleontologists can find them. Finally, these deposits must have chemical signatures, created by the steady radioactive decay of elements, that allow them to be aged. The fossil record is an incomplete history of life on Earth, but human understanding of evolutionary history would be impoverished without it.

Origins

Life began in primeval seas under an atmosphere of ammonia, methane, and hydrogen that would be deadly today (Oparin, 1936). How did organic molecules form from this deadly inorganic broth? Stanley Miller (1953) solved this riddle when he created amino acids, the building blocks of proteins, by subjecting hydrogen, methane, ammonia, and water vapor to electrical discharges that simulated lightning. Creation of sugars, polypeptides, and nucleotides is now commonplace under simulated conditions of the Earth's early atmosphere.

In fact, almost any form of excess energy, including volcanic eruptions, ultraviolet radiation, and lightning, is sufficient to create organic molecules out of surface materials that were common on the primeval Earth.

What were the first self-replicating entities capable of evolution? Most experts suspect that simple nucleic acids formed first, and then, through spontaneous chemical bonding, assembled polypeptides that compartmentalized and catalyzed primitive cellular functions (Eigen and Schuster, 1982). Other scientists implicate polypeptide bubbles, called protenoid microspheres, which form easily in the proper medium, expand like simple organisms, and divide (Fox and Dose, 1972). The proteins might then have assembled nucleic acids capable of both replication and production of new protein molecules (Dyson, 1982). The difficulty in both scenarios is to imagine how either nucleic acids or polypeptides could accurately control their own division into unique units. Thomas Cech (1985) may have solved this quandary by showing that protozoan rRNA can modify its activation energy, thereby controlling its own splicing into unique units. Primeval nucleic acids may not have needed *efficient* protein catalysts. Once formed by whatever route, rudimentary cells with self-replicating nucleic acids would have been subjected to natural selection (Rokhsar et al., 1986). Families of primitive assemblies of proteins and nucleic acids might then have differentially reproduced, with those best able to exploit the organic soup crowding out less efficient assemblies.

Early evolution was immeasurably slow. Traces of impressions of simple microscopic organisms exist in some Precambrian rocks that are 3.5 billion years old. It took another 2.5 billion years for recognizable invertebrate body plans to appear (Table 9-1). The earliest Precambrian organisms were **heterotrophs** that assimilated organic compounds in their environment. These were soon followed by single-celled forms that could use energy from the sun to catalyze metabolic reactions. These first **autotrophs,** forerunners of primitive photosynthesizers, were undoubtedly engulfed by new kinds of heterotrophs that were capable of feeding on the primitive autotrophic organisms. Herbivory as we know it began in Precambrian seas as soon as multicellular invertebrates could eat and digest multicellular algae. Marine and aquatic grazers were undoubtedly well advanced by the time that the first organisms found refuge on land.

Early Life on Land

The first true land plants, with primitive vascular tissues similar to those of modern liverworts, sprawled across damp hollows 440 million years ago, rooted in a thin film of wet soil formed by algal colonists. Within a few million years, better-developed vascular systems allowed arborescent (treelike) plants to stand up and compete for space and light more effectively (Stewart, 1983). Early communities were dominated by Psilopsida, soon replaced by ancestors of present-day club mosses (Lycopsida), horsetails (Equisitales), and ferns (Pteridophyta). All of these early vascular plants were dependent on standing water for repro-

Fig. 9-2 Fern life cycle. For 300 million years of their 440 million year history, vascular plants relied on flagellated male gametes that required a film of water to swim to the egg. Relicts of this once universal mode of reproduction are found in ferns and other primitive vascular plants. (A) sporophyte; (B) sporangia; (C) spore; (D) gametophyte (prothallus); (E) reproductive organs; (F) new sporophyte. After Barth (1985).

duction (Fig. 9-2). Alternation of spore-producing (sporophyte) and gamete-producing (gametophyte) generations can still be traced from its algal origins to a much reduced form in flowering plants. In early land plants, alternation of generations separated a large competitive stage, the sporophyte, from a minute stage that produced sessile eggs and flagellated male gametes that had to swim through a film of water to reach the egg. The gametophyte stage preserved the water-dependent reproduction characteristic of algal ancestors.

Early land plants probably never had the Silurian and Devonian mud and sun to themselves. Plant-eating millipedes came ashore with them in the Silurian, and wingless insects evolved early in the Devonian. By the late Devonian, winged insects and amphibians that ate them were common.

The chemical arms race between herbivores and plants can be traced right to the beginning of life on land. Early land communities harbored pathogenic fungi and plant-eating insects that challenged the defenses of primitive vascular plants (Swain, 1978). The first defenses were probably structural. Most plant groups produced virtually indigestible cellulose and hemicellulose skeletons, and the evolution of silica skeletons made horsetails almost inedible to plant-

Table 9-2 Evolution of defensive chemicals in plants[a,b]

Class of Compound	Defense Against			Plant Dominance (millions of years ago)				
	Path	Ins	Vert	Psilotum (400)	Horsetail (370)	Fern (320)	Gym (200)	Ang (70)
Phenols								
Simple	+	−	(+)	+	+	+	+	+
Tannins	+	+	+	−	(+)	+ +	+ +	+ +
Isoflavenoids	+	−	−	−	−	−	−	+
Terpenoids								
Mono-	−	+	+	−	−	−	+ +	+ +
Sesqui- and Di-	+	+	+	−	−	+	+ +	+ +
Tri-	−	+	+	−	−	+	+ +	+ +
N-containing								
Nonprotein								
amino acids	+	+	+	−	−	+	(+)	+
Alkaloids	(+)	+	+	−	−	−	(+)	+ +
Cyanogens	−	+	+	−	−	(+)	(+)	+
Glucosinolates	−	+	+	−	−	−	−	+

Source: After Swain (1978).

[a]Under enemies of plants, (+) or (−) means presence or absence of defensive capability. Under plant class, signs indicate the relative importance of the indicated secondary chemical class. Brackets indicate a tentative result.

[b]Path, pathogens (bacteria, fungi, viruses); Ins, insect; Vert, vertebrate; Gym, gymnosperm; Ang, angiosperm.

eating animals. A more elaborate evolution of chemical defenses began early in the history of life on land (Table 9-2). In particular, the introduction of phenolic and terpenoid secondary compounds in the first 100 million years of land life simultaneously protected plants against pathogens and arthropod herbivores.

Evolving insect orders and the rise of plant-eating reptiles placed ever increasing challenges on plant defenses (Swain, 1978). At first, condensed tannins and terpenoid insect hormone mimics, the ecdysones, were the prominent secondary defenses. Reptiles, relatively unaffected by condensed tannins or insect hormone analogues, stimulated the evolution of hydrolyzable tannins and other allelochemicals that were effective against both vertebrates and rapidly evolving insects. Vast coal beds and oil reserves dating from the Carboniferous testify to the effectiveness of plant defenses for millions of years. Herbivores and detritivores, beset with rapidly proliferating chemical defenses in vegetation, could not keep up with plant productivity.

As plant stems and foliage became less and less palatable to herbivores, plant reproductive parts became increasingly attractive. Thickened spore-producing organs on early vascular plants suggest that insects fed on spores, and may have dispersed them. The first definite evidence of something like pollination by insects comes from pollen-like spores of seed ferns in the family Medullosaceae in the upper Carboniferous (Crepet, 1983). These grains, up to 0.6 mm wide, are far too large to have been effectively carried by wind. Predacious

Fig. 9-3 Fleshy fruitlike seeds occurred as early as the Jurassic in early seed plants like *Caytonia nathorsti*. In this specimen lower seeds had been shed. These may have been among the first reptile-dispersed plants. After Sporne (1965).

insects, relatives of modern lacewings (Neuroptera), probably transported these spores by accident as they clamored over the still separate male and female plants in search of prey 300 million years ago. By the Age of the Dinosaurs 190 million years ago, cycads had evolved the hermaphroditic condition with male and female sexual parts on the same plant. These cycads, probably fertilized with the help of beetles similar to those alive today, were closely related to the ancestors of true flowering plants.

Seed dispersal by animals also has a venerable heritage. As early as 200 million years ago, Jurassic progenitors of modern cycads bore fleshy fruitlike seeds that appear to have been adapted for consumption by herbivorous reptiles (Fig. 9-3; Sporne, 1965). Dispersal of these ancient gymnosperm seeds by reptiles was probably analagous to contemporary seed dissemination by large herbivorous mammals that digest most seeds that they eat, but fail to kill them all. Echoes of this ancient phenomenon still exist on the Galapagos Islands, where giant tortoises eat fruits and disperse some seeds in viable condition (Rick and Bowman, 1961). Cycads such as *Macrozamia* of the Australian sand plains still produce fleshy seeds like those that the dinosaurs ate, but the seeds are now scattered by oppossums (Burbidge and Whelan, 1982). The land areas of the Earth now belong to the angiosperms.

Rise of the Angiosperms

To Charles Darwin, the origin of the angiosperms was an "abominable mystery." Angiosperms, or true flowering plants, have been such a puzzle since Darwin's day because many taxa, now recorded as fossils, appeared simulta-

Table 9-3 Competitive advances of angiosperms over gymnosperms

1. In angiosperms, sterile and fertile appendages are aggregated into a **flower,** which is often showy, scented, and stocked with nectar for attracting pollinators. In gymnosperms, sterile and fertile appendages are usually separated and pollination is usually by wind.

2. In angiosperms, **juxtaposed micro- and megasporophylls** permit efficient pollen transfer, reducing the amount of pollen needed. In gymnosperms, pollen- and ovule-producing structures are separated, necessitating massive pollen production.

3. In angiosperms, **double fertilization** of the egg and polyploid endosperm permits the entire unit to be aborted with little loss of investment if pollination fails. In gymnosperms, seeds mature before fertilization, leading to substantial waste if fertilization fails.

4. In angiosperms, the **closed carpel** (ovary) protects ovules against desiccation and herbivores, and provides a surface (stigma) and route (style) for pollen capture and pollen tube growth. In gymnosperms, ovules are less protected and pollen capture is precarious.

5. In angiosperms, **reiterated tissues** permit replacement of damaged tissues with the original growth model of the tree or shrub. In gymnosperms, replacement is impossible.

neously in almost modern form at the beginning of the Cretaceous, 135 million years ago. The constellation of characters that distinguished them from more primitive vascular plants seems to have gelled in a very few million years (Table 9-3). An absence of transitional forms between lycopsids, seed ferns, and cycads and the dominant plants on Earth has been a challenge to paleobotanists for well over a century. Where did the flowering plants come from? Once evolved, why did they diversify and dominate the land areas of the Earth so quickly?

The mystery of their origins is by no means solved, but modern botanists have more clues than Darwin did. G. Ledyard Stebbins (1974) pointed out that the pollen of living primitive angiosperms decays easily and is carried by insects. For both reasons, it rarely finds its way into the lake and pond deposits that record the history of wind-pollinated species. The same factors may account for the rarity of pollen fossils in Mesozoic rocks. Moreover, the earliest angiosperm pollen fossils now known from the Jurassic are similar to spore fossils. Fossil angiosperm pollen grains may exist in Mesozoic rocks, but lie undetected among fossil spores.

But why do fossil stems and leaves appear in the early Cretaceous, and not before? Stebbins and others argue that early flowering plants probably did not live in the lowland Jurassic swamps that have yielded so many fossils of ferns and cycads. The first natural experiments in angiosperm evolution may have been shrubs or trees of arid highlands where stems and leaves rarely fossilize. Two key angiosperm traits (Table 9-3), pollen that is independent of standing water and a closed protective carpel around the ovary, may be adaptations for drought resistance. Actual fossils will probably be lost until a paleontologist happens on a shallow pool sediment or a tufa (calcium carbonate) secretion, left on some Jurassic mountain, that by chance trapped ancient leaves and somehow escaped erosion for 150 million years.

The **arid-origin hypothesis** makes subsequent events more plausible. The

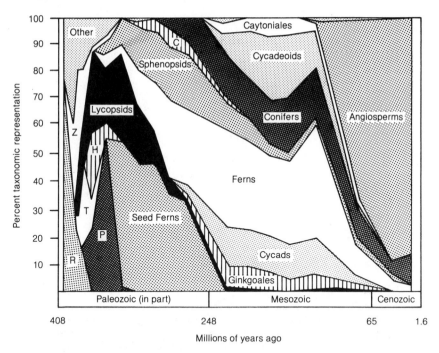

Fig. 9-4 Taxonomic composition of subtropical and tropical lowland floras. Note that angiosperms had virtually replaced all other taxa by the early Pleistocene, 1.6 million years ago. Ancient taxa referred to by letters include Cordaitales (C), Hyeniales (H), Progymnosperms (P), Rhyniophytes (R), Trimerophytes (T), and Zosterophyllophytes (Z). After Knoll (1986).

dawn of the Cretaceous was marked by a global drying and cooling of the climate, resulting in vast environmental changes that led to the total extinction of the dinosaurs and widespread extinctions of insects and marine life. The reasons for this climatic change are hotly disputed by geologists. Some implicate a massive meteor strike (Alvarez et al., 1980), others postulate unprecedented volcanic activity (Officer and Drake, 1984). Both scenarios argue that huge clouds of dust and ash blocked the sun, directly killing most plants and the herbivores that depended on them. Ensuing climatic changes over the entire globe altered the conditions of life for the survivors. Angiosperms and gymnosperms, long adapted for drought by virtue of pollen and flower morphology and seeds capable of dormancy, were strongly favored over more primitive plants dependent on standing water for fertilization. Climatic changes that eliminated serious competition from primitive vascular plants marked the first stage of the takeover of the Earth by flowering plants.

Once started, the angiosperm revolution gathered force through the Cretaceous, practically eliminating the more primitive gymnosperms from all but marginal high altitude and high latitude environments (Fig. 9-4). Both basic ecological factors and key improvements in morphological design (Table 9-3) made the angiosperm revolution inexorable.

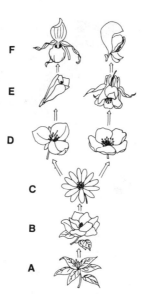

Fig. 9-5 Trends in flower shape over 100 million years. (A) The earliest flowers had no discernible symmetry. (B) Flower of open hemispherical shape without clear symmetry. (C) Typical open radially symmetrical flower (yellow adonis). Subsequent divergence often altered shape in monocots (left) and dicots (right). (D) Reduced but fixed number of flower parts in the spiderwort (left) and buttercup (right). (E) Hidden nectar and bilateral symmetry in the freesia (left) and columbine (right). (F) Complex and strongly zygomorphic flower shapes in the lady's slipper orchid (left) and monkshood (right). From Barth (1985); after Leppik (1972).

Pollination and dispersal mutualisms gave flowering plants an edge over their wind-pollinated and wind-dispersed gymnosperm rivals. Philip Regal (1977) argued that wind pollination is only effective at close quarters, when plants such as conifers grow in virtually monospecific stands. Wind-blown pollen is seriously impeded if even one or two trees of other species stand between flowering pines or spruces. Insect-pollinated plants, on the other hand, easily set seed when growing among other species because their pollinators seek out isolated individuals. Regal argued that bird- and mammal-dispersed angiosperms diluted gymnosperm forests with colonists that blocked wind-blown pollen. The flowering plants that impeded gymnosperm reproduction were not at a disadvantage, because their flowers could attract pollinators from substantial distances. To Regal (1982) it is no coincidence that communities dominated by conifers and recently evolved wind-pollinated angiosperms persist in high latitudes, on mountains, and in other climatic situations where insect pollination is unreliable. Elsewhere, they simply cannot compete against plants that are assisted in their reproduction by animal pollinators.

Finally, progressive adaptations in morphology and development gave the angiosperm ground plan a decided edge over gymnosperms (Table 9-3). Cretaceous angiosperms were not closely related to modern forms, but fossil pollen and seeds show that modern genera and, in some cases, species evolved by the middle and late Tertiary (Doyle and Hickey, 1976, Tiffney, 1977). Accompanying the taxonomic evolution were changes in floral efficiency that transformed simple amorphous flower design into the variety of specialized and highly efficient designs known today (Fig. 9-5). Perhaps as important was the evolution of the basic angiosperm developmental design of reiteration in growth tissues, or meristems (Tiffney, 1981). The new angiosperm design of reiterative tissue growth allowed these plants to regenerate a normal growth habit if branches or their meristems were destroyed by physical damage or herbivores. Under most conditions in a mixed forest of competing gymnosperms and angiosperms, the flowering plants held decided advantages in both reproductive and vegetative attributes.

Animal Radiations

Primitive birds and mammals also existed during the Jurassic, but they were completely dominated in size, number, and importance by reptiles. *Archeopteryx,* the first known bird, had both avian and reptilian features. True feathers identified it as a bird, but teeth and a long bony tail indicated its close reptilian ties. The earliest mammals were small insectivores that resembled present-day shrews. The key adaptive breakthrough shared by these primitive creatures was homeothermy, the ability to maintain body temperature and consequently activity during the night chill or spells of cold weather. This was not as great an advantage in warm Jurassic forests as it would soon be during and after the Cretaceous extinctions.

Climatic events that destroyed the fern and horsetail forests and their dinosaur inhabitants left an open field for hitherto insignificant angiosperms, birds, and mammals. Insects had already diversified by the Cretaceous, although many, such as the social bees that are familiar to us today, radiated well after the angiosperm explosion began. However, almost all of of the modern orders of birds and mammals appeared or underwent major adaptive radiations during the Cretaceous and the first 20 million years of the Tertiary. The herbivorous mammal fauna is a well-known example of taxonomic diversification that paralleled plant diversification.

RECENT HISTORY: THE LAST 20 MILLION YEARS

The objective of this section is to explore puzzling paleoecological questions and their implications for interpreting modern communities. The historical issue concerns evolution and extinction of assemblages of large mammals, the megafauna, in North America. Extensive fossil remains document an abundant and diverse megafauna over the last 20 million years. Virtually all of these large mammals are extinct. One cannot but wonder what it was about earlier worlds that favored these huge creatures, nor why they are gone. The historical record of the rise and fall of the megafauna is a window into pre-Pleistocene America.

Mass extinctions in the past leave puzzling questions for ecologists. How should modern communities be interpreted in light of the past? Did the vegetation so crucial to these large plant-eaters change, forcing most to extinction? In that case contemporary prairies and forests might represent communities that impoverished and ultimately destroyed the megafauna. Or did something else kill the mammals, leaving plant communities behind? In that event ecologists might now be studying incomplete relics of the past—anachronistic plant assemblages without the mammalian selective agents that shaped them.

Forest, Savanna, Steppe, and Glacier

Twenty million years ago North American forests and savannas witnessed the burgeoning of an enormous fauna of enormous mammals, the mammalian megafauna (Kurten and Anderson, 1980). The peak of familial and ordinal mammalian diversity was reached at about this time, and the proportion of *large* mammals (>5 kg) was on the rise (Bourlière, 1975). A bestiary of the time would have included many animals, including antelopes, camels, elephants, and horses, that we now associate with Africa and Asia (Figure 9-6; see Anderson, 1984). Other huge mammals, such as the giant beaver and several species of giant ground sloths, left only diminutive relatives. Peccaries, bison, tapirs, and deer are relics of the past that give some inkling of what the Miocene megafauna must have been like.

Why did the megafauna evolve? Large body size was favored in response to changes in climate that displaced forests with vast savannas over the mid-

Fig. 9-6   Large mammals of North America during the Tertiary included relatives of modern horses, camels, rhinoceroses, elephants, and antelopes now restricted to Africa and Asia. Shown here on the left are short-legged rhinoceroses *(Teleoceras)* from Kansas. On the right are primitive mastodonts *(Trilophodon)*, showing tusks on both the upper and lower jaws. After a mural by Charles Knight, courtesy of Chicago Field Museum of Natural History.

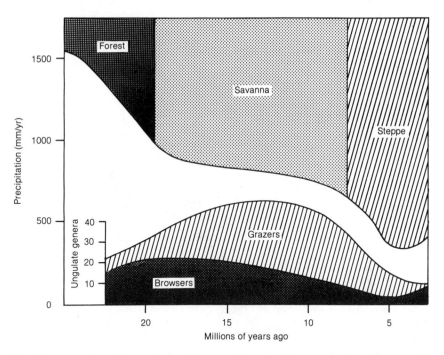

Fig. 9-7 Increasing aridity during the Miocene (19 to 5 million years ago) produced a shift from predominately forest to savanna to predominately steppe (top scale). This shift in habitat was accompanied by an increase in the proportion of grazers to browsers among fossil ungulates (bottom scale). Lowered productivity on arid steppes led to a decline in both browsing and grazing ungulates 3–5 million years ago. After Webb (1983).

continent (Figure 9-7). Large size offers a refuge from predation, especially in open country. Even the Miocene complement of pumas, lions, and hyenas could not kill large mammals with impunity (Bakker, 1983). Moreover, the mechanics of locomotion make it much easier for large rather than small ungulates to conserve energy and water as they forage between widely scattered water holes or feeding grounds (Pennycuick, 1979). Between 20 and 10 million years ago, North America became a vast mosaic of forest remnants populated by browsing sloths, deer, and elephant-like gomphotheres and mastodonts, interspersed with limitless savannas teeming with wild horses, antelope, and mammoths. This was a world quite like that of East Africa today, but even more diverse.

Why did the megafauna disappear? Most of the answer is inherent in the long-term climatic changes that displaced forests with long-grass savannas, and finally with arid steppes (Webb, 1983). As forest diminished, the nutritional niches of browsers disappeared. Fifteen million years ago there were seven genera of North American horses, four of which ate broad-leaved plants (Figure 9-8). Five million years later only 3 of 12 genera were browsers. Soon, even these were gone. Long-term habitat changes from forest to savanna, and finally from savanna to much less productive steppe, explain in a broad sense why

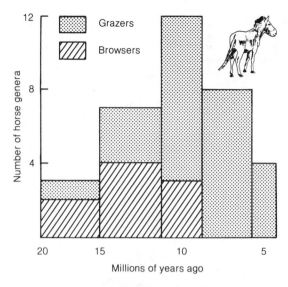

Fig. 9-8 Horse genera of the Miocene of North America. Horses underwent extensive adaptive radiation, reaching a maximum of 13 genera 10 million years ago and falling to 4 genera 5 million years ago. Consistent with general trends (Fig. 9-7), the dental features of several early genera indicate a diet of leaves and buds, while those of later genera suggest a diet of grasses. After Webb (1983).

many genera of large mammals vanished. Those adapted to eat relatively protein-rich, but chemically defended, broad-leaved plants were constricted to smaller and smaller ranges. Later, the grazers that were dependent on luxuriant savannas failed to compete on arid steppes.

One of the most intriguing facets of the megafaunal extinctions is not *that* they happened, but *how* they happened. A scenario outlined above might predict steady depletion and eventual extinction of one taxonomic group after another. This is not what happened. Sharp increases and decreases in the number of horse genera from one period to another suggest bursts of adaptive radiation followed by episodes of extinction. A broader view confirms episodic extinction for all mammal genera (Figure 9-9). The time resolution of radiocarbon dating decreases as one goes back into the past, perhaps inflating the length of time over which the extinctions occurred for the earlier intervals. But it is clear that at least eight episodes of extinction occurred, with some of massive proportions associated with some of the 22 glaciations that gripped North America during the Pleistocene (Graham, 1986). For instance in the Late Hemphillian, 5 million years ago, 62 mammal genera vanished. Thirty-five members of the megafauna were among them.

Most controversy surrounds the last extinction episode, perhaps because more fossils are available for animals that vanished only 8,000–15,000 years ago than for those that disappeared millions of years ago. Forty-three genera and many more species of mammals vanished in a time span that may have been as short as a few decades, or as long as 5,000 years. What might have caused

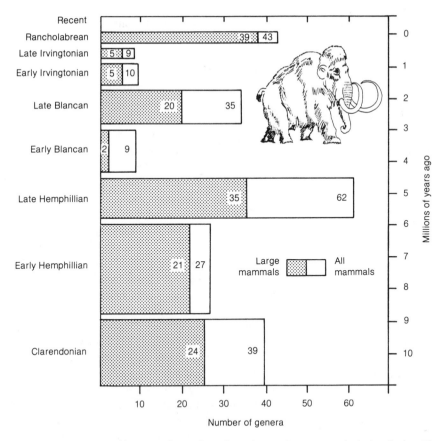

Fig. 9-9 Extinctions of large (≥5 kg) and small North American mammals during the last 10 million years. Shown are periods of extinction associated with distinctive geological formations (left axis). Note that the largest extinction, during the Late Hemphillian 5 million years ago, included as many small as large mammals. A greater proportion of large mammals became extinct in the most recent episode, during the Rancholabrean between 35,000 and 10,000 years ago. Extinction episodes appear shorter in more recent episodes because the definition of the fossil record is better. After Webb (1984).

such abrupt devastation? Might lessons from this last event throw light on previous extinction episodes?

The last megafaunal extinction coincided with the recession of the Wisconsin glaciers. Many paleoecologists favor the **environmental deterioration hypothesis** (Graham, 1986), which suggests that withdrawal of the glacial ice sheets set in motion climatic changes that disrupted the feeding and reproductive ecologies of the large mammals.

How might glacial withdrawals disrupt megafaunal ecology? R. Dale Guthrie (1984) suspects that a mosaic of swamp, forest, savanna, and steppe that was characteristic of most of the Pleistocene was displaced by the extensive monotonous floral associations that we consider normal today. A high diversity of browsers forced into a smaller and smaller variety of forest habitats lost many species, as did a high diversity of grazers forced into monotonous prairie and steppe habitats. Deer, elk, and moose were the winners in the first competition, bison and pronghorn antelope in the second.

Fossils show ample evidence of ecological disruption. Bones of mammals that survived the glacial retreat were of normal size, but antlers, horns, and tusks of megafauna that became extinct were dwarfed, suggesting widespread malnutrition among animals that were soon to become extinct (Guthrie, 1984). Postglacial communities were chaotic from a contemporary perspective. Margaret Davis (1976, 1986) found mixtures of plant species in pollen samples in eastern North America unlike any known today, and further noted that records for previous interglacials often differed in this respect from each other. Similarly, animal species occurred in mixtures of northern and southern species that are unknown today (Guthrie, 1984). Pleistocene spruce grouse *(Canachites canadensis)* and bog lemmings *(Synaptomys cooperi),* now animals of the far north, mixed with cotton rats *(Sigmodon hispidus)* and hog-nosed skunks *(Conopatus leuconotus)* that are now subtropical or tropical in distribution. One implication of dwarfism and chaotic species associations is that Pleistocene communities were in tremendous ecological flux. Large mammals with large appetites were probably at a disadvantage in a world of constantly changing food supplies and competitive environments.

An alternative scenario exists. Paul Martin (1967, 1983) favors the **human overkill hypothesis,** which holds that Asian hunters crossed the Bering Straits and swept across North and South America, slaughtering camels, horses, mastodons, and other large mammals in their paths. Pleistocene hunters may have exterminated some survivors of previous episodes of extinction. Ample evidence of ecological disruption and the near absence of the remains of extinct species in the campfire sites of early hunters argue against a crucial human role (Guthrie, 1984; Graham, 1986). The Asian immigrants witnessed the demise of the megafauna, but they probably did not cause it.

Plants the Megafauna Left Behind?

Sixty million years of natural selection by large herbivorous mammals undoubtedly shaped the evolution of many plant characteristics. Might the well-armored cacti of the American deserts be echoes of previous millenia when ground sloths, camels, and native wild horses were an everpresent threat to plants? Could bulky anomalous fruits with no apparent means of dispersal be botanical anachronisms, adapted to a fauna of dispersal agents long extinct?

Extinct mammalian herbivores almost certainly shaped the evolution of cacti in the southwestern United States (Janzen, 1986). Alternative hypotheses that would explain formidable cactus armor that is clearly directed at mammals (Fig. 3-2A) are not obvious. Wild peccaries *(Tayassu tajaeu)* sometimes eat cacti (Fig. 3-1A), but they prefer seeds, roots, and tubers (Sowls, 1984). Moreover, all but the heaviest grazing by domestic cattle, goats, and horses favors proliferation of cacti, often converting arid grassland to cactus desert. Plants of spiny desert and thorny scrub probably were strongly favored during millions

5 cm

Fig. 9-10   Real and suspected megafaunal fruits. Left: *Drypetes gossweileri* (Euphorbiaeceae), an elephant fruit from Gabon, West Africa. Right: *Gustavia superba* (Lecythidaceae), a bulky Central American fruit, now scattered by rodents, that might once have been eaten by extinct megafauna of the pre-Pleistocene. Drawings by R. Roseman from photographs by L. Emmons and H. F. Howe.

of years of evolution in North America, much as they are today in arid African thornbush in which the dominant herbivores are large mammals.

Daniel Janzen and Paul Martin (1982) further pose the **megafaunal fruit hypothesis,** which argues that many living temperate and tropical fruits are botanical anachronisms, adapted for dispersal by an extinct megafauna. Some African fruits dispersed by elephants bear a remarkable resemblance to fruits of tropical American plants that often rot directly under the trees that produced them (Fig. 9-10). The Central American *Gustavia superba* (Lecythidaceae), illustrated here, is an especially likely candidate because its seeds and seedlings survive well in dense aggregations near their parents (Sork, 1985). Chemical adaptations for survival in dense bunches in animal dung might have saved it from extinction when the megafauna vanished. For many other fruits first attributed to the syndrome, the case for megafaunal dispersal is more controversial (Howe, 1985). Bats, monkeys, or rodents disperse many, calling into question the need to hypothesize dispersal agents in the past. Seeds and seedlings of other common tree species virtually never survive insect or pathogen attack near their parent trees. These species must have living dispersal agents to have survived 10,000 years without the megafauna, but their dispersal ecology has not yet been studied.

SUMMARY

Dramatic and ecologically important interactions between plants and animals date from the beginning of multicellular life. Among the early adaptations of plants invading the land were indigestible cellulose, hemicellulose, and silica, followed closely by such secondary compounds as the tannins, terpenoids, and nonprotein amino acids that are still with us today. Spore transport by predatory insects and propagule transport by reptiles date back at least 200 million years. The evolutionary relationships that set the stage for contemporary life are pervasive in the history of plant and animal adaptation.

The modern world of plant and animal relationships began with the environmental changes that ended the Cretaceous period, and with it the reign of the dinosaurs, seed ferns, and gymnosperms. These animals and plants, dominant for hundreds of millions of years, were quickly replaced by warm-blooded birds and mammals that were capable of withstanding wide swings in temperature, and by drought-adapted angiosperms that depended on insects for pollination and vertebrates for seed dispersal. Most contemporary patterns of herbivory, plant defense, pollination, and seed dispersal evolved in the first few million years of the Tertiary period.

Better records in younger rocks show that plant populations in North America were profoundly influenced by both long-term and short-term climatic changes. These in turn influenced herbivore evolution and extinction. A rich fauna of large mammalian herbivores of the Miocene diminished over a span of 10 million years as increasing aridity in turn favored forest, forest and savanna, savanna, and unproductive steppe. These changes in turn favored browsers, a diverse mixed fauna of browsers and grazers, grazers, and then a small minority of survivors to the present day. Punctuating this overall trend were numerous glacial events, each lasting at most tens of thousands of years, that catalyzed the evolution of some mammalian groups and precipitated the extinction of others.

Much of the controversy over the last Pleistocene extinctions exists because of the concordance of two events tied to glaciation. The glaciers locked up so much water that the Bering land bridge emerged from the sea and admitted human immigrants to North America. The same glaciers set in motion environmental changes that profoundly altered relations between plant communities and herbivores, quite apart from human influence. What is known for sure is that North American pre-Pleistocene plant and animal communities were quite unlike any known today, and that most of their large mammalian inhabitants disappeared with them.

The extinction of the megafauna leaves ecologists with the puzzle of interpreting the world that the huge animals left behind. Extensive cactus deserts, fostered by overgrazing by domestic cattle and horses, are reminders of times when millions of plant-eating camels, horses, and cattle roamed North America. Anomalous tropical fruits, resembling species fed upon by elephants in Africa and Asia, may also be echoes of the previous world of native mammalian herbivores. The communities themselves, however, are gone.

STUDY QUESTIONS

1. Is it safe to assume that the first multicellular animals fed on autotrophs? Why or why not?

2. What preconditions in the alternation of generations would be necessary for primitive spore dispersal by insects to be useful to plants?
3. Some paleobotanists think that the angiosperm ovary evolved as protection against insects that fed on flower parts, while others see it as an adaptation for preventing desiccation. How could one test these two hypotheses?
4. Sketch a scenario that would account for the demise of the gymnosperm forests in favor of animal-pollinated and animal-dispersed angiosperms.
5. Megafaunal assemblages all but disappeared as climatic changes turned North American habitat mosaics into monotonous savannas and steppes. Why did this not happen in Africa, which retains extensive savanna and steppe as well as much of its megafauna?

SUGGESTED READING

Wilson Stewart (1983) gives an authoritative up-to-date account of early plant evolution in *Paleobotany and the Evolution of Plants*. The evolution of flowering plants is discussed by G. Ledyard Stebbins (1974), *Flowering Plants: Evolution Above the Species Level*, and in an anthology edited by Charles Beck (1976), *Origin and Early Evolution of the Angiosperms*. Alfred Sherwood Romer's (1959) classic, *The Vertebrate Story*, is a pleasant and informative sketch of vertebrate evolution. All sides of the controversy over causes of megafaunal extinctions are discussed in *Quaternary Extinctions: A Prehistoric Revolution*, edited by Paul S. Martin and Robert G. Klein (1984).

10

Plants and Animals
in Modern Communities

Do plant and animal relationships actually influence the distribution and abundance of species? Fossil evidence suggests that plant communities have profoundly influenced the number and character of herbivorous animals over evolutionary time. The historical record also indicates that plant-eating insects, reptiles, and mammals have, through their ability or inability to overcome plant defenses, shaped plant communities. This chapter considers interactions between contemporary plants and animals that influence community composition through their influence on the distribution and abundance of species.

The amount of energy available for plant and animal growth and metabolism is constrained by the physical limitations of an ecosystem. The trophic pyramid (Fig. 1-3) illustrates energy accumulation and loss at each step of the food chain, from the point of energy capture by plants to plant consumption by herbivores and then to herbivore consumption by carnivores. Energy loss to growth, metabolism, and inefficient digestion make it inevitable that herbs are more common than rabbits, and that rabbits are more common than foxes. Simple trophic considerations cannot, however, explain species composition *within* a trophic level. The pyramid gives no clues as to why some herbs are more common than others, or why rabbits are more or less common than muskrats. The goal of community ecology is to understand the interactions and historical events that determine species abundances in any given ocean, forest, pond, or meadow.

Many factors determine whether a species will be common, rare, or altogether missing in a natural community. Physiological constraints play a role. Neither tropical trees nor the insects that eat their leaves could survive cold Minnesota winters, nor could Minnesota oaks grow on thin tropical soils. History is also important. Cold-adapted trees of the South American Andes might survive in Minnesota, but simply never got there. Of concern here are the relationships among species that influence species abundances. Ecologists have traditionally seen competition and predation as the key interactions that shape natural communities, but this exclusive emphasis is rapidly changing. Recent studies show that herbivory profoundly influences the success of species in some natural communities, and that mutualisms play a vital role in others.

Studies of plant and animal relationships are opening ecological frontiers that were unimagined a few years ago.

This chapter will probe ways in which plant and animal relationships influence the distribution and abundance of species, and consequently influence the composition of natural communities. Examples from the literature of herbivory and mutualism illustrate how plant and animal relationships *directly* influence the ability of species to survive and reproduce, or *indirectly* shift the competitive balance among species.

THEORY

The most useful approach for analyzing interactions between species has been solution of simultaneous equations that specify the effects of growth of each of two species on each other. Gause and Witt (1935) used **Lotka–Volterra equations** to explore the effects of competition on the population growth of two species. For two species i and j:

$$dN_i/dt = r_i N_i[(K_i - N_i - a_{ij}N_j)/K_i]$$
$$dN_j/dt = r_j N_j[(K_j - N_j - a_{ji}N_i)/K_j]$$

where N_i and N_j are population sizes of species i and j, t is time, r_i and r_j are intrinsic rates of natural increase for species i and j (calculated as birth rate minus death rate; see Chapter 1), and K_i and K_j are the carrying capacities of species i and j (the numbers of individuals of each species that a habitat can support). Perhaps unfamiliar are the interaction coefficients a_{ij} and a_{ji}. When two species compete for the same resource (e.g., food), a_{ij} is the effect of an individual of species j on the growth of population i. Likewise, a_{ji} is the effect of population growth of an individual of species i on the growth of species j.

The Lotka–Volterra equations can be applied to any ecological relationship in which the presence of one species influences the population size of another. Competitive interactions between species depress the fitness of each species, conveniently represented by $-/-$, where each negative sign represents the depressive effect of one population on the growth of the other. Gause and Witt realized that a_{ij} and a_{ji} need not be negative; they can be positive or zero (Table 10-1). This suggests at least six common *classes* of relationships between species. Furthermore, the positive or negative effects signified by a_{ij} and a_{ji} should be continuous because they can be large or small in magnitude. Mistletoes, for instance, might boost bird populations only a little (small $+ a_{ij}$, where the bird is i and the plant is j), but seed dispersal by birds might be critical for successful mistletoe reproduction (large $+ a_{ji}$).

Mutualisms offer an instructive example of the application of community theory to plant and animal relationships (Addicott, 1981; Wolin, 1985). Lotka–Volterra theory describes the interactions of species i and j by the general equations:

Table 10-1 Fitness effects of ecological relationships between species[a]

0/0	Neutralism: Species interact without affecting each other's fitness.
0/+	Commensalism: One species gains with no affect on the other.
0/−	Amensalism: One species suffers, the other does not.
−/−	Competition: Two species use the same limiting resource(s).
−/+	Herbivory, parasitism, predation: One species eats another.
+/+	Mutualism: Two species benefit.

[a]For each of two interacting species, signs show that a relationship increases fitness (+), decreases fitness (−), or has no effect (0).

$$dN_i/dt = N_i f_i(N_i, N_j)$$
$$dN_j/dt = N_j f_j(N_i, N_j)$$

where f_i and f_j are the functions that describe how growth rates (r_i and r_j) and interaction coefficients (a_{ij} and a_{ji}) relate to each other. Carole Wolin (1985) reviews a variety of algebraic forms that the f function may take. For purposes of illustration, one plausible form is

$$f_i = (r_i + a_{ij}N_j)(1 - N_i/K_i)$$

In other words, the growth of population i depends on its intrinsic rate of increase plus the enhancement of population j due to j's effect on i, all discounted by the relative abundance of i. This f_i function then can be substituted in the general equation for dN_i/dt above. A parallel function f_j would be created for species j for substitution in dN_j/dt.

This example illustrates the per capita benefit of mutualism, for the benefit to i decreases as its own density increases. This is what Wolin calls a "low-density mutualism" because it limits the benefit of mutualism if, for instance, a large patch of flowers or stand of fruiting trees satiates pollinators or dispersal agents. "High density" and "density-independent" models further describe stable mutualisms in which the benefit increases with the density of the recipient, and even describe a few stable instances in which benefit is independent of the density of the recipient.

The role of theory is to sort out the consequences of important interactions in nature, not to mimic nature. The result of model construction should be a predictive theory, based on a reasonable understanding of key biological variables. Predictive theory guides observation and experiment. The process of refining the simplest models to better approximate nature is a valuable learning experience. For instance, realistic models of stable mutualisms suggest reasons to explore features of communities—such as flower or bee density, or the effects of plant densities on herbivore densities—that were simply not perceived as interesting a few years ago.

LIMITS ON POPULATIONS: DIRECT EFFECTS

Animals can kill or assist plants, and thereby influence plant distribution and abundance. Plants can likewise provide a crucial resource for animals, and by their presence or absence make survival possible or impossible for herbivores, pollinators, dispersal agents, or protectors.

Animal Influences on Plants

Some animals consistently influence plant abundances and spatial dispersions, either by killing seeds or plants outright or by providing a critical service. In other cases the influence of animals is not consistent, but occurs periodically when drought or other climatic stresses force animals to alter feeding habits.

Seed Predation in Goldenbush

Goldenbush (*Haplopappus squarrosus*) presents a paradox in southern California. The species is uncommon near the Pacific coast of California and is abundant inland. Physiological constraints on growth or supression by superior competitors near the coast would be traditional explanations for such a distribution. The puzzling feature of goldenbush natural history is that uncommon plants near the coast are large, apparently healthy, and flower profusely. They show no signs of physiological stress or supression. Based on the number of flowers produced per plant, seed production and seedling recruitment should be greater near the coast than inland. Ultimately, goldenbush fecundity would pre-

Fig. 10-1   Observed and expected frequencies of goldenbush *(Haplopappus squarrosus)* on a roadside transect from California coastal scrub to interior mountains. Expected frequency is calculated from the total number of flowers initiated per individual plant. Observed frequency approximates abundance (density × frequency) in 15,250 sites at 167-m intervals from coast to interior. After Louda (1982a).

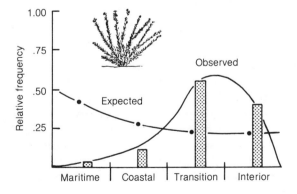

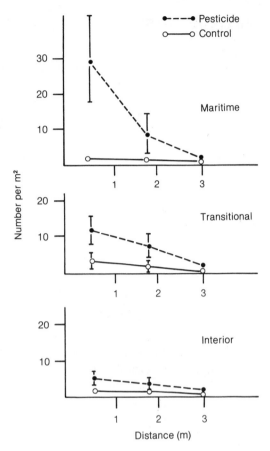

Fig. 10-2 Seedling density near goldenbushes *(Haplopappus squarrosus)* with and without insects excluded by insecticide. Solid lines indicate control plants not treated with insecticide; dashed lines indicate insecticide treatment. After Louda (1982a).

dict much higher adult populations near the coast than inland (Fig. 10-1), precisely the opposite of the observed distribution.

Svata Louda (1982a,b) explored this paradox with an elegant series of experiments. First she noticed that flower heads were infested with larvae or adults of several species of thrips, flies, moths, and wasps that kill flowers and developing seeds. To test the hypothesis that these seed predators reduced seed production, Louda compared the fecundity of plants treated with insecticide-laden water with plants treated only with water in the coastal scrub of California (Louda, 1982b). Insecticide treatment reduced fruit loss to insects from 94 to 41%. Insecticide treatment of adult plants did not increase the chances that individual seedlings would survive. But because more seeds survived in treated than control study plots, seedling recruitment was higher in treated than control plots. Might high levels of insect infestation near the coast limit goldenbush abundance there?

Having established a cause of low seed production on the coast, Louda tested the hypothesis that insect infestation had a greater effect on the coast than inland (Louda, 1982a). Insecticide treatments at three sites confirmed that insects influenced seed production and ultimately seedling recruitment more in maritime vegetation on the coast than inland (Fig. 10-2). Large flower heads of coastal plants might have attracted more insects than smaller flower heads in the interior (see the section on resource concentration below), or more amenable climates might have favored higher insect species diversity near the coast than inland. Whichever the reason, it is clear that goldenbush in maritime habitats is strongly limited by flower and seed destruction rather than by physiological stress or competition.

Wood Rats and Saguaro Cacti

The Sonoran Desert of Arizona and Mexico is a surreal world of bare earth, spiny shrubs, bushlike cacti, and the towering treelike saguaro cactus *(Carnegiea gigantea)* (Fig. 10-3A). Saguaro cacti are usually widely scattered throughout the desert (Fig. 10-3C). One obvious hypothesis that would explain this dispersed spatial pattern is that saguaros compete with each other and with other plants for a limiting resource: water. Desert communities *are* water-limited, with precipitation rarely reaching 300 mm each year and potential evapotranspiration (water loss from evaporation and transpiration) exceeding 2,000 mm. It is certainly plausible to suppose that saguaros do not grow close together because dry desert soils do not provide enough water to support aggregations of these giant neighbors. Without conflicting evidence, the most parsimonious interpretation of saguaro spacing is competition for water.

But should ecologists *assume* that competition controls saguaro dispersion? The natural history of the saguaro helps frame the testable hypotheses (Steenbergh and Lowe, 1977). Fruit-eating birds disseminate 40 to 60 million saguaro seeds per hectare (100×100 m) each year, of which 160,000 to 240,000 (16 to 24 per m^2) germinate. Virtually all of these die of freezing and drought within the first year, with no obvious influence of competition. Saguaros might still compete at much reduced densities, but massive density-*in*dependent mortality is bound to reduce the effects of a density-*de*pendent process, such as competition. The question is, "Do the seedlings that survive extremes of hot and cold compete, or do other factors determine their dispersion?"

Raymond Turner and his colleagues (1969) suspected that intraspecific competition was an insufficient answer because saguaros sometimes do grow in pairs or clumps. Interspecific competition for water was also suspect, because many saguaros grow in close contact with other large plants (Fig. 10-3B). If water were the key limiting factor, neither could occur. What might explain the scattered distribution of large saguaro cacti?

Saguaro seedlings die on open bare earth, but live in the shade. This runs counter to what one would expect if seedlings competed with established plants. Competition should be highest close to established cacti or shrubs that have substantial needs for water, and lowest away from established plants in the

Fig. 10-3 Saguaro cactus *(Carnegiea gigantea)* of the Sonoran Desert of Arizona and Mexico. (A) Old adult. Note that no young saguaro cacti are nearby. (B) Juvenile saguaro displacing its former nurse tree, a thorny palo verde tree *(Cercidium microphyllum)*. (C) Healthy stand of cacti of medium age. Photographs by H. F. Howe.

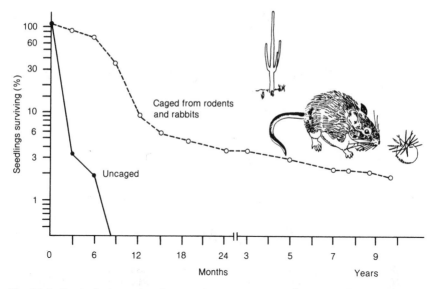

Fig. 10-4 Survival of caged and uncaged saguaro cacti seedlings transplanted in Saguaro National Monument, Arizona. Cages excluded rodents and rabbits that eat the seedlings. All 800 uncaged seedlings were dead within 1 year; 1.9% of the 800 caged seedlings survived at least 10 years. Data from Turner et al. (1969).

open. Turner and his co-workers found that unshaded seedlings were killed directly by the intense desert sun, before they had an opportunity to compete for water or anything else.

Furthermore, seedlings survived best in the shade of other cacti or spiny trees like palo verde *(Cercidium microphyllum)*, rather than in the shade of less formidable plants (Fig. 10-3B). This suggested the hypothesis that spiny "nurse plants" protect seedlings from wood rats *(Neotoma albigula)* and other herbivores. Caged seedlings in the shade survived far better than uncaged seedlings in the shade, showing that herbivores are a key source of saguaro seedling mortality (Fig. 10-4). The saguaro story demonstrates that what seems obvious is not always true. Far from excluding young saguaros through competition, other plants offer necessary protection from desiccation and herbivory.

Hyraxes and Giant Lobelias

Herbivory may be a continual challenge to plants, or it may occur only periodically under exceptional ecological circumstances. African hyraxes and giant lobelias of tropical alpine mountains offer an excellent example of the importance of unusual events in plant ecology.

Like many tropical alpine communities, the upper slopes of Mount Kenya are dominated by giant forms of genera that are usually small herbs (Fig. 10-5). The giant treelike groundsel *Senecio keniodendron* is in the same family (Asteraceae) as goldenbushes in California. Similarly, the giant lobelias *Lobe-*

Fig. 10-5 Rock hyrax *(Procavia johnstoni)* herbivory on giant lobelia *(Lobelia telekii)* in alpine habitats of Mount Kenya, East Africa. These rodent-like mammals are actually related to elephants. The lobelias, which stand 2 m high, form dense forests at elevations of 3,500 to 5,000 m. Usually hyraxes and lobelias coexist with little interaction. During years of drought when alternative foods disappear, however, hyraxes decimate lobelia populations. Periodic devastation by hyraxes may convert lobelia-covered slopes to grassland. Drawing by Todd Erickson.

lia telekii and *Lobelia keniensis* resemble garden lobelias in flower form, but are 2 m high. This strange community of the tropical highlands offers two examples of the influence of vertebrate herbivory on community composition.

First, elephants *(Loxodonta africana)* periodically invade the valleys to elevations of 4,000 m, feeding extensively on the giant plants and sometimes converting these unique forests into grasslands (Mulkey et al., 1984). Records do not document the time scale required to fully understand these invasions, but it is clear that at least once every few decades the *Senecio* and *Lobelia* forests below 4,000 m are destroyed by elephants.

Second, the higher alpine zone above 4,000 m suffers periodic destruction by smaller herbivores. Elephants do not venture to the upper reaches of Mount Kenya, but herbivory by a rodent-like mammal, the rock hyrax *(Procavia johnstoni)* is a continual but usually minor source of mortality on giant lobelias (Young, 1984; Fig. 10-5). Tough leaves and a poisonous latex protect giant lobelias from most herbivores, but hyraxes often supplement their broad diets with lobelia leaves. Under normal climatic conditions, hyraxes are a minor source of mortality for the plants.

Long-term demographic studies of the giant plant communities of Mount Kenya fortuitously documented a rare event. A demography of *Lobelia telekii* begun in 1977 by Truman Young showed healthy populations of this species, characterized by numerous individuals and low mortality in all age classes (Young, 1984). A multi-year drought intervened after 1980, drastically decreasing the availability of other hyrax food plants. Hyraxes devastated *Lobelia* plants growing near their colonies. At different study sites on the mountain, hyraxes accounted for 77–100% of the deaths of plants over 4 years of drought (Young and Smith, 1987). Perhaps most intriguing, hyraxes selectively killed slow-growing individuals that probably had defenses reduced by drought stress (Young, 1985). During the first year of drought alone, 36% of the plants with growth rates below the population mean were killed, whereas 23% of those growing faster than the average were killed. After 4 years, drought and hyrax herbivory eliminated all but the healthiest lobelia individuals near hyrax colonies.

Hummingbirds and Tropical Flowers

Does the absence of a mutualist actually alter the ability of its partner to survive or reproduce? Studies of pollination, seed dispersal, and ant protection imply mutual dependence of plants and animals, but the removal experiments that would clearly demonstrate the fact are still in their infancy. Islands with and without certain hummingbird species offer natural experiments for evaluating the importance of mutualists to each other.

Trinidad and Tobago are two tropical islands near the Caribbean coast of South America. Trinidad (4520 km^2) has 34 species of hummingbird-pollinated plants and 11 species of hummingbirds. Tobago (295 km^2) has 31 hummingbird-pollinated plants and only 5 hummingbird species (Feinsinger et al., 1985). Both islands are occupied by two plant species that have very different requirements for pollination by hummingbirds. *Justicia secunda* (Acanthaceae) is a large herb with flowers accessible to virtually any hummingbird. Both long and short-billed species can reach its nectar. *Mandevilla hirsuta* (Apocynaceae) is a vine with a deep corolla accessible only to hummingbird species with long bills. Of special interest is the fact that both long- and short-billed hummingbirds are common on Trinidad, but only short-billed species are common on Tobago.

Yan Linhart and Peter Feinsinger (1980) tested the hypothesis that the reduction of pollinator diversity affects specialist plants more than generalist plants. They reasoned that *Justicia* would suffer less on Tobago than *Mandevilla*, because *Justicia* can be pollinated by virtually any hummingbird. *Mandevilla*, on the other hand, might be expected to have difficulty reproducing on small islands like Tobago, where long-billed hummingbirds are rare.

Direct observation lent some support to the hypothesis. *Justicia* was visited as much on Tobago as on Trinidad, while *Mandevilla* was visited less often on the smaller island. Most intriguing were the effects of rich and depauperate pollinator communities on pollen movement, as estimated by dispersal of colored dye from flowers of each species. Only a minor reduction in dye dispersion occurred for *Justicia* on Tobago, but far less dye was carried from *Mandevilla* on Tobago than on Trinidad (Fig. 10-6) suggesting that *Mandevilla* suffered reduced outcrossing on Tobago. Dye dispersal is an imperfect measure of pollen dispersal (Thomson et al., 1986), but independent evidence also sup-

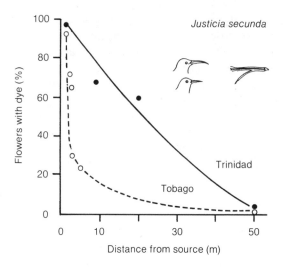

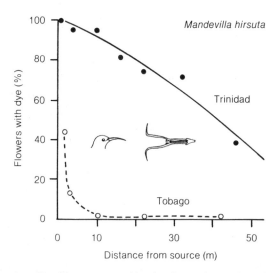

Fig. 10-6 Pollination efficiency as estimated by dye dispersal away from flowers on Trinidad (solid circles), which has 11 species of hummingbirds, as compared with Tobago (hollow circles), which has 5 species of hummingbirds. *Justicia secunda,* a generalist, suffers little from a reduction in the hummingbird fauna. *Mandevilla hirsuta,* a specialist, suffers drastically reduced pollen dispersal on Tobago, where long-billed hummingbirds are rare. After Linhart and Feinsinger (1980).

ports the contention that specialists suffer in depauperate communities. The reduction in the potential for outbreeding by *Mandevilla* vines on Tobago was compounded by a lower seed set (3%) on Tobago than on Trinidad (6%). As predicted, the specialist flower suffered more from a reduced mutualist assemblage than did the generalist flower.

Overview: Animal Effects On Plants

The four examples considered above illustrate the potential effects of animal feeding habits on plant abundances and distributions. Consistent herbivory like that documented for goldenbush flowers and saguaro seedlings influences the distribution, abundance, and spatial pattern of plant species. Such effects are probably widespread, but undocumented. Hyrax destruction of giant lobelias on Mount Kenya during drought is one of many examples of the influence of mammalian herbivory during climatic cycles or changes of seasons. Similar interactions of drought and diet choice by elephants and ungulates occur in African savannas (see Sinclair and Norton-Griffiths, 1979). Finally, *Mandevilla* dependence on long-billed hummingbirds for pollen transport illustrates the potential for dependencies among mutualists. As discussed below under "keystone mutualists," such dependencies may prove to be of the utmost importance in diverse tropical communities.

Plant Effects on Animals

Plants influence animal populations by inducing migration, enhancing survival, and affecting reproduction of animals that use them for food. Two examples illustrate how plants influence herbivore populations.

Resource Concentration Hypothesis

A general observation in agricultural and natural communities is that herbivores tend to seek out rich food sources. Richard Root (1973) formalized this observation as the **resource concentration hypothesis.** The hypothesis predicts that specialist insect herbivores will favor high densities of host plants over low densities because they will find and breed on dense patches more easily than on patches with few plants. An ancillary hypothesis, the **enemies hypothesis,** predicts that parasitoids should control herbivores more effectively in mixed than in pure stands of hosts. The two hypotheses are important for ecologists because herbivores in nature must usually select among host stands that contain varying numbers of host and nonhost species.

The resource concentration hypothesis is deceptively difficult to test. Catherine Bach (1980) opted for an experimental design to tease apart the effects of host density and plant species diversity on beetle population dynamics. Bach planted cucumbers *(Cucumis sativus)* at high (289/100 m^2) and low (144/100 m^2) densities in monocultures and in polycultures that also contained corn *(Zea mays)* and broccoli *(Brassica oleracea).* She then monitored colonization and reproduction of a specialist herbivore, the striped cucumber beetle *(Acalymma vittata).* The results were striking (Fig. 10-7). Host plant density, by itself, had little influence on beetle density. Moreover, there were no appreciable differences in parasitism on herbivores in these different treatments, so the enemies hypothesis failed. What influenced beetle populations far more than host den-

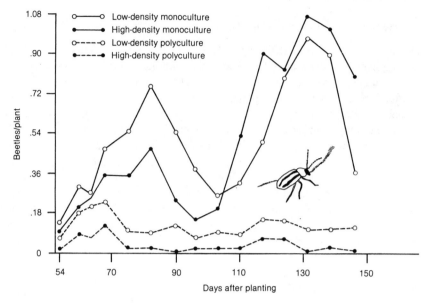

Fig. 10-7 Mean number of striped cucumber beetles *(Acalymma vittata)* on cucumber plants *(Cucumis sativus)* in garden plots in monocultures, or in polycultures planted with corn *(Zea mays)* and broccoli *(Brassica oleracea).* Density of cucumber plants (289 or 144/100 m^2), by itself, has little influence on beetle density. (●———●) Monoculture high density; (○———○) monoculture low density; (● - - - ●) polyculture high density; (○ - - - ○) polyculture low density. After Bach (1980).

sity was the presence or absence of plants (corn or broccoli) *other than* the cucumber host species.

Why are nonhost species more important to leaf beetles than host density? Bach suspects that shading by taller species made cucumbers in polycultures less palatable than those in sunny monocultures, in which all plants were approximately the same size and open to the sun. Plants in monocultures were in fact much larger and more productive in all measures than those shaded by corn. Ironically, cucumbers did not live as long in monocultures as in polycultures. *Acalymma* herbivory is by itself negligible, but the beetle transmits a bacterial wilt disease *(Erwinia tracheiphila)* that kills many plants when beetle density is high. In an annual plant, this has little effect on fruit production because the disease strikes late. Disease transmission in dense patches would have a much more important effect on long-lived perennial plants.

Can the resource concentration hypothesis be accepted or rejected on the basis of present evidence? As Bach and others point out, the prediction that insect herbivores should find or stay longer in patches of high density is confounded in agricultural experiments in which several insect herbivores increase in total numbers per plant at high densities (Kareiva, 1983). The question cannot be answered at present because so few studies simultaneously manipulate host density and other relevant variables, such as species diversity of nonhost plants. Host colonization is an important factor in insect population biology, but the factors controlling insect movements are not yet fully understood.

Small Mammal Cycles

Plant quantity and quality influence animal population dynamics by affecting the reproduction of generalist herbivores. For instance, browse (twigs and foliage) quantity and quality influence the numbers of ungulates or rodents in grassland and forest communities. Population cycles of small mammals offer an example of unusual historical interest.

Fur records of the Hudson's Bay Company trading posts suggest mammalian predator/prey cycles (Fig. 10-8). Canadian populations of the snowshoe or varying hare *(Lepus americanus)* appear to show 10-year cycles of boom and bust in population size. For nearly a century, ecologists used this as a classic example of a prey population cycle driven by predation. Lynx *(Lynx canadensis)* populations grow as their prey populations grow. Eventually, so the predation story goes, the cats become so numerous that they depress hare populations, sending the latter into a steep decline that ultimately causes starvation among the predators.

Causation should not be inferred from association. Associated lynx and hare cycles fit nicely into the expectations of predator–prey theory, so lynxes were thought to cause hare cycles. As Michael Gilpin (1973) irreverently pointed out, an unbiased observer of the data might as easily have concluded that hares eat lynx! Evidence now suggests that lynx cycles are simply tied to the abundance of food animals; the cats become common as their prey prosper and

Fig. 10-8 Lynx and hare cycles recorded by furs traded to the Hudson Bay Company. Originally, hare cycles were thought to be caused by lynx predation. Now it is known that deterioration of food quantity and quality initiates the hare decline, bringing down the lynx population with it. Herbivore damage induces food plants to produce adventitious shoots that are heavily defended with toxic terpenes and resins, thereby further reducing the quality of already depleted supplies of food plants during peaks of hare abundance. Data from Elton and Nicholson (1942).

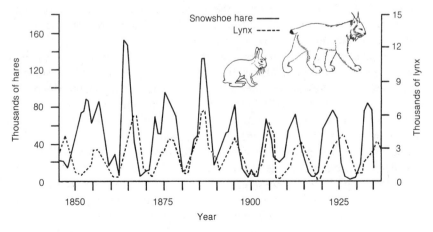

Table 10-2 Comparison of the predation and food limitation interpretations of the lynx and hare population cycles; interpretations are identical, except for items 1, 3, and 4

Predation Hypothesis	Food Shortage Hypothesis
1. **Hare populations increase due to low predator populations.**	1. **Hare populations increase due to food abundance.**
2. Lynx populations keep pace with hare populations.	2. Lynx populations keep pace with hare populations.
3. **Lynxes become so numerous that they depress hare populations.**	3. **Hare browse declines in quantity and nutritional content; toxic allelochemicals increase.**
4. **Hare numbers decline due to intense lynx predation.**	4. **Hare numbers decline due to starvation; lower fecundity.**
5. Lynxes starve; population crashes.	5. Lynxes starve; population crashes.
6. Hare populations recover.	6. Hare populations recover.

starve when their prey starve. There is little evidence that lynx ever eat enough hares to *cause* a population crash in their prey. What now appears to drive the hare cycles is the quantity and quality of their own food, not predation.

Snowshoe hares live in a severe world of bitter winter cold and deep snow. Each day of the long subarctic winter, a hare requires 3000 g of woody browse less than 4 mm in diameter from several plant species, including willow *(Salix spp.)*, aspen *(Populus tremuloides)*, rose *(Rosa acicularis)*, hazel *(Corylus cornuta)*, saskatoon *(Amelanchier alnifolia)*, and scrub birch *(Betula glandulosa)* (Pease et al., 1979). Field enclosures in Alberta monitored by Pease and his co-workers for six consecutive winters demonstrated that there simply was not sufficient browse available during peak years in the hare cycle to support densities of up to 8 hares per hectare. Heavy browsing leading up to the peak years had dramatically reduced the protein, fat, and carbohydrate contents of key food plants, further reducing the capacity of northern forests to support peak populations of these voracious herbivores. Moreover, the animals required several suitable food species to survive the winter. Captives fed on only one species died, regardless of the quantity of food given. Starvation, not lynx predation, initiated population crashes in snowshoe hares.

A far clearer picture of the reasons for hare cycles exists now than was possible even a decade ago (Table 10-2). Lloyd Keith (1983) and his students have found that hares at high densities suffer weight loss, sharply increased winter mortality, and markedly lower spring reproduction among the survivors. Reduction in the nutritional content of food plants during years of heavy browsing is now known to be complemented by active defenses by the plants. Alaskan scientists have shown that alders and birches contain a variety of toxic secondary compounds that deter hare feeding (Bryant et al., 1985). Perhaps most intriguing, intense browsing results in production of secondary shoots heavily defended with terpenes and phenolic resins that are extremely unpalatable to hares (Bryant, 1981). Intense browsing results in chemical defense of plants already depleted in quantity and nutritional quality by herbivory.

Overview: Plant Effects on Animals

Plants clearly influence animal numbers. The influence of plant dispersion on animal numbers and the feedback of herbivore-induced changes in plant chemistry on animal populations are two general issues of central importance in community ecology. Plant dispersion and abundance now figure importantly in models of mutualism (Wolin, 1985), and are likely to increase in importance in the effort to understand herbivore population dynamics. Herbivore-induced plant defenses could have profound effects on insect as well as mammal population dynamics (Haukioja, 1980). The feedback between herbivore numbers, damage to forage plants, chemical defenses, and response by herbivores is one of the most exciting areas of current research in the ecology of herbivory.

COMMUNITY EFFECTS

To what degree do strong ecological interactions between pairs or guilds of plants and animals account for differences among natural communities? Are the immigrations, extinctions, and different reproductive successes of organisms that underlie differences in community composition dependent on interactions between species, or independent of them (e.g., MacArthur and Wilson, 1967)? The answers are not known in depth because few ecologists have tried to distinguish these fundamental alternatives. Recent studies of plant and animal relationships do suggest that some interactions profoundly influence community composition.

Grazing Succession

Natural grasslands once supported vast herds of large herbivorous mammals and birds over millions of square kilometers of the Earth. East Africa now harbors the last extensive remnants of ecosystems that once covered much of North and South America, Asia, and Australia (Fig. 10-9). In microcosm, the Serengeti-Mara Plains of Tanzania and Kenya offer a living laboratory for the study of processes that probably were of fundamental importance throughout the world before the advent of modern civilization.

Herbivory is a governing feature of the Serengeti-Mara ecosystem. Most plant communities suffer a consistent but moderate level of herbivory, amounting to less than 10% of the plant biomass produced each year. Over 90% of the grass produced in some parts of the Serengti Plains is consumed by large mammalian herbivores each year. How can several species of large plant-eating mammals live on such a heavily used resource?

The answer lies in ungulate diets and distributions. Larger rumen or intestinal volumes allow buffalo, zebra, and large antelopes to eat a much wider variety of plants than smaller, and necessarily more selective, antelopes. Furthermore, tremendous geographical and seasonal variability in plant species

Fig. 10-9 African grasslands. Top: Kongoni, a large antelope, on a tall-grass savanna near Nairobi, Kenya. Note the tall senescent grasses. Bottom: Zebra on the lawn-like Serengeti-Mara plains of southwestern Kenya. Photographs by H. F. Howe.

composition and growth provide different grazers with different diets (McNaughton, 1985). Seasonal and annual variations in climate that alternatively favor some grazers over others help to maintain herbivore diversity.

Given differences in diets of small and large ungulates, might some grazers

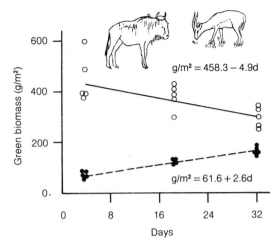

Fig. 10-10 Effects of wildebeest grazing on vegetation in the western Serengeti plains. The biomass of green vegetation decreases as grasses become senescent in fenced plots protected from wildebeest grazing (open circles). Green vegetation is rejuvenated after migratory wildebeest herds pass through (solid circles). Young green shoots stimulated by large mammal grazing are especially palatable and accessible to smaller herbivores, such as Thomson's gazelle. After McNaughton (1976). Copyright © 1976 American Association for the Advancement of Science.

actually facilitate feeding by other species? D. F. Vesey-FitzGerald (1960) noticed that grazing by large herbivores seemed to maintain grass swards (stands) in a perpetual state of youth, allowing smaller antelopes to feed on highly nutritious young shoots. Large animals initiated a chain of feeding bouts by successively smaller species in a process called **grazing succession.** In support of this idea, R. H. V. Bell (1970) noted that some Serengeti ungulates tended to occur together in pairs of species of very different sizes. Topi (*Damaliscus korrigum;* 90–140 kg), for instance, often fed with eland (*Taurotragus oryx;* 450–700 kg) or with gazelles (*Gazella granti;* 42–68 kg), not with animals their own size. Might large herbivores actually *stimulate* grass growth by eating old silicified grasses and thereby help smaller species?

Samuel J. McNaughton (1976) tested the grazing succession hypothesis directly for one grazing association. McNaughton fenced portions of the grasslands dominated by the grasses *Themeda triandra* and *Pennisetum mezianum* from enormous herds of wildebeest (*Connochaetes taurinus;* 200–228 kg). Unfenced plots suffered what appeared to be a devasting reduction in green biomass (457 to 69 g/100 cm^2), height (66 to 29 cm), and biomass concentration (6.9 to 2.4 mg/10 cm^3). Subsequent observations showed that the devastation was more apparent than real. Over the following month, the grazed portion of the sward increased production at a rate of 2.6 g/100 cm^2/day, while the ungrazed fenced plots *decreased* production at a rate of 4.9 g/100 cm^2/day. Moreover, groups of Thomson's gazelles (*Gazella thomsonii;* 18–25 kg) fed on the ungrazed plots at a rate of 0.27 g/100 cm^2/day, but fed on the grazed areas at the much higher rate of 1.05 g/100 cm^2/day (Fig. 10-10). Ungrazed swards were senescent, while areas grazed by larger herbivores produced younger and more palatable shoots.

Might these data suggest a mutualistic relationship between wildebeest and Serengeti grasses (Owen, 1980)? McNaughton (1985) thinks not. Compensation in biomass production does not necessarily indicate a higher reproductive success among eaten as compared with uneaten grasses. Moreover, heavy grazing clearly puts some grasses at a disadvantage. Grazing promotes the spread of unpalatable species that discourage herbivores (McNaughton, 1978, 1985). Within species, transplant experiments show that heavy grazing selects for higher silica content, lower stature, and asexual tiller production (McNaughton and Tarrants, 1983; McNaughton, 1984). Grasses evolve ways of circumventing herbivore damage, not inducing it. In a general theoretical scheme (Table 10-1), $-/+$ interactions characterize the relationships between wildebeest and grasses and $0/+$ those between wildebeest and gazelles, at least at some times and places.

Keystone Species

Keystone species are animal or plant species with a pervasive influence on community composition. Removal or extinction of keystone species profoundly changes the competitive relationships, and consequently the relative abundances, of other species in a community. Keystone species may be so common that they use a disproportionate share of resources. If a competitive keystone disappears, other plants or animals that play similar roles in the community prosper. Alternatively, keystones may directly or indirectly influence species in trophic levels other than their own. For instance, Robert Paine (1966) discovered that experimental removal of the predatory starfish *Pisaster ochraceous* allowed the mussel *Mytilus californianus* to monopolize space that, in the presence of the starfish, harbored 11 species of limpets, clams, and mussels. Keystone predators have now been identified in a variety of rocky intertidal communities of tropical and temperate coasts.

Mutualists may also be keystone species (Gilbert, 1980). A **keystone mutualist,** typically a plant, provides critical food resources for pollinators or fruit-eating animals that play important roles in the reproduction of other plants at other times of the year. Both practical and moral considerations preclude experimental removal of suspected keystone mutualists in many communities, but an examination of one observational study suggests how the keystone concept may be applied to mutualism.

A canopy tree of the Costa Rican rainforest, *Casearia corymbosa*, is a keystone mutualist in the complex tropical forest of Finca La Selva, a research station operated by the Organization for Tropical Studies (Howe, 1977). This uncommon tree bears a heavy crop of small oily arillate fruits during a 10-week season that includes December, the month of lowest fruit production in the La Selva forest (Fig. 10-11). Seedlings that germinate under parent trees face rapid density-dependent mortality, as compared with those that fall a few

Fig. 10-11 Birds that feed on fruits of the keystone mutualist, *Casearia corymbosa*, in the rainforest of eastern Costa Rica. In the foreground is a masked tityra *(Tityra semifasciata)*, a reliable dispersal agent for this tree. In the background, a large territorial toucan *(Ramphastos swainsonii)* attempts to chase other birds away from a fruit-laden tree. Because *Casearia corymbosa* bears fruits during a period of fruit scarcity in the forest, it supports at least six species of birds that benefit it little, but nonetheless disperse seeds of many other trees and shrubs at other times of the year. Drawing by R. Roseman.

meters away. One species of bird, the masked tityra *(Titrya semifasciata)*, consistently disperses the seeds throughout the *Casearia* fruiting season. Twenty-one other bird species also eat the fruits, but are not especially reliable dispersal agents for the tree. Among these are fruit thieves, like the parrot *Amazona autumnalis*, that strip off the edible arils and drop the seeds without dispersing them. Frugivorous flycatchers *(Myiozetetes similis* and *Myiozetetes granadensis)* ignore the tree early in the season, but depend on it to supplement an insect diet when the January dry season depresses insect numbers. Still others, like the toucan *Ramphastos swainsonii*, only feed on *Casearia* when their preferred foods are unavailable during December scarcity. Even though they eat prodigious quantities of *Casearia* seeds, toucans are poor dispersal agents for this tree. Some individual toucans simply occupy a *Casearia* crown day and night and drive off other birds. As soon as preferred fruits appear in January or February (e.g., *Virola koschnyii)*, toucans abandon *Casearia. Casearia* not only provides food for its reliable dispersal agent, the masked tityra, it also supports a variety of opportunists through periods of food scarcity.

What would happen if either *Casearia corymbosa* or its principal dispersal agent, the masked tityra, were exterminated from La Selva? Enough is known to make some educated guesses (Howe, 1984b). The disappearance of a key food during annual fruit scarcity would quickly push the tityra, toucans, and some other birds to local extinction. They appear to have little else to eat in December. The masked tityra disperses the seeds of many tropical trees at other times of the year. Disappearance of this bird would slowly erode the competitive position of some of these plants. Toucans are dispersal agents of large-seeded fruiting trees, such as *Virola surinamensis,* which was discussed earlier (Howe et al., 1985). Extinction or dramatic reduction in numbers of toucans at La Selva would eventually spell doom for *Virola koschnyii* and other large-seeded trees normally dispersed by these giants of the forest canopy. Could a widening circle of extinctions extend to other birds that eat fruits normally dispersed by toucans or tityras? No one knows. The ecological ramifications of this keystone mutualism have never been investigated.

Seasonality accentuates the importance of keystone mutualisms. This is particularly true for small fruit-eating birds that quickly reach an energy deficit when food is scarce (Worthington, 1982), but fruit-eating mammals are also susceptible to local famine. Robin Foster (1982a,b) studied the seasonality of fruit fall in the seasonal forest on Barro Colorado Island, Panama. As at La Selva, December was a critical period of fruit scarcity. In 1970, irregular rains during the normal January to March dry season disrupted tree flowering and caused a massive fruit failure of 40 common tree species, including 12 staples of this predominately mammalian frugivore assemblage (e.g., *Dipteryx, Chrysophyllum, Brosimum, Tetragastris, Spondias.*) Instead of a month-long shortfall in fruit, scarcity extended from July 1970 to March 1971. Foster found unprecedented numbers of dead howler monkeys, agoutis, coatis, peccaries, white-faced monkeys, and other fruit-eating animals on the forest floor. An animal carcass occurred every 300 m of trail. Populations were large enough on Barro Colorado Island to weather the famine, and these mammal species have now recovered. In small isolated reserves, the outcome probably would have been less fortunate. Unusual weather could spell disaster for small populations of animals dependent on keystone mutualists.

White-Sand Forests

Biotic relationships exist within the context of a physical world that may itself influence the number, identity, and ecological properties of plants and animals that it harbors. Perhaps it is fitting to close with the example of soil attributes that profoundly affect the relationships between plants and the animals that eat their leaves, pollinate their flowers, or disperse their seeds.

Lowland tropical forests are noted for a diverse and dense vegetation, poor soils, high rainfall, and rapid ecological succession where treefalls, landslides, or other disturbances temporarily expose the soil. Tropical forests differ widely in ecological characteristics (Vitousek and Sanford, 1986). The total amount of plant matter in mature forest varies from 6 to 80 kg/m^2 (Whitaker 1975). The lower end of the scale includes scrub forests growing on nutrient-deficient white-sand soils (Fig. 10-12). Such forests, called caatinga in South America, may be small habitat islands of one or a few hectares or cover hundreds to thousands of square kilometers. These are associations of short drought-adapted trees and shrubs in which up to 60% of the total plant biomass is in roots, many of which form a mat across the surface of the ground (Anderson, 1981). Some caatinga

A

B

Fig. 10-12 Stunted forests on white-sand soils in South America. Caatinga scrub (top) is from central Amazonia, Brazil; white-sand heath (bottom) is from San Carlos de Rio Negro, Venezuela. Forests on these nutrient-poor soils may be limited to one or a few hectares, or cover thousands of square kilometers. Plant communities on these impoverished soils have tough, heavily defended foliage that supports far fewer animal species than the foliage of nearby forests on richer soils. Photograph of caatinga scrub from Anderson (1981); photograph of white-sand heath by Carl F. Jordan, University of Georgia.

plants contend with extraordinary nitrogen and phosphrous shortages by catch-
ing leaf litter as it falls—before it hits the ground! Heavy cutinization protects
caatinga plants from seasonal drought, replacing the deciduous habit characteristic of many nearby trees on more nutrient-rich soils.

Serious nutrient shortage for plants has consequences for animals. Daniel
Janzen (1974) reasoned that plants limited by nutrient deficiency have difficulty
replacing leaves. Trees and shrubs growing on nutrient-poor soils should be
evergreen, while those a few meters away on rich soils are likely to be deciduous during drought. Nutrient stress also may affect the ability of a plant to
replace leaves eaten by herbivores. A tree growing next to a caatinga may be
able to replace leaves eaten by a monkey, while a tree growing on the white
sands often lacks the nutrients to recover from such damage. How might these
nutrient limitations influence animal communities?

Janzen reasons that forests on white-sand soils should be characterized by
(1) tough foilage heavily defended by chemicals, (2) minimal herbivory, (3)
low numbers and biomass of herbivores, and (4) extremely rare carnivores at
the top of the food chain because of a paucity of primary consumers (herbivores). Indirect tests support these predictions. White-sand forests are noto-
riously lacking in bird, mammal, and insect life, suggesting a depauperate herbivore community at the base of the consumer chain (Fig. 1-3). Moreover,
black acidic rivers flowing from extensive white-sand basins are loaded with
humic acids (tannins and other phenolic acids) and bits of undecomposed plant
matter, suggesting extraordinary secondary defenses.

The diets of leaf-eating monkeys offer a direct test of the hypothesis that
plants invest in defenses in proportion to the cost of replacing lost tissues.
Doyle McKey and his co-workers (1978, 1981) compared diets of colobus
monkeys *(Colobus satanas)* in the white-sand forests of the Douala-Edea Re-
serve in Cameroon, West Africa with the diets of close relatives *(Colobus bad-
ius* and *Colobus guereza)* feeding in forests on richer soils in the Kibale Forest
of Uganda, Central Africa. As predicted, the leaves that were available to monkeys in white-sand forests were far less nutritious, and far more heavily defended, than those growing on richer soils in Uganda (Table 10-3). Moreover,
monkeys in the Cameroon forest avoided most common trees, feeding selectively on rare deciduous trees and on uncommon herbaceous vines. Only 37%
of the food of *Colobus satanus* consisted of leaves; 53% consisted of seeds. In
contrast, monkeys of the richer Uganda forest obtained at least 75% of their
food from leaves of common trees and ate proportionately fewer seeds. Perhaps
because suitable food was so scarce in the white-sand forest, only a tenth as
many *Colobus* lived in the Cameroon site as in the much richer Ugandan site.

The example of white-sand forest ecology both ties together earlier concepts,
and suggests as yet untested predictions relevant to communities of interacting
plants and animals. From the perspective of plant apparency to animals, all but
rare deciduous trees with periodic leaf flushes and herbaceous vines on white-
sand soils are heavily defended with secondary chemicals (see Feeny, 1976).
From the perspective of plant energy budgets, these nitrogen- and phosphorus-
poor soils support plants with carbon-based defenses of phenols (see Coley et

Table 10-3 Composition, energetic value, and chemical defenses of mature leaves of common trees in a forest on white-sand soils (Cameroon, West Africa) and lateritic soils (Uganda, Central Africa)[a]

Assays	Lateritic Soils Mean	White-Sand Soils Mean
Potential Food Value		
Percentage ash	12.1	3.7
Percentage nitrogen	2.8	1.7
Percentage phosphorus	0.2	0.1
Gross energy (cal/g)	4,204.3	4,833.9
Defensive Properties		
Total phenols (mg/g)	35.2	75.7
Condensed tannin (mg/g)	25.5	46.4
Realized Food Value		
Percentage digestibility[b]	29.5	13.7
Gross energy × percentage digestibility	1,215.3	653.1

Source: Data from McKey et al. (1978).

[a]Lateritic soils are weathered aluminum- and iron-rich soils.

[b]Digestibility is the percentage of dry matter digested after 96 hours in cow rumen inoculum.

al., 1985). Further questions await attention. A low number and biomass of animals should make pollination, and especially seed dispersal, far more problematic on poor than rich soils. One would predict a much higher proportion of wind-pollinated and wind-dispersed trees in the white-sand forest than in richer forests. Constraints on plant nutrition should be expected to influence their palatability, and therefore the distribution and abundance of animals that eat their leaves, pollinate their flowers, and disperse their seeds.

SUMMARY

Plant and animal relationships influence the species composition of entire communities. Herbivores can limit plant populations directly, as illustrated by insect destruction of goldenbush flowers and by rodent consumption of saguaro seedlings. To the degree that these plants influence the shrubland or desert communities where they live, herbivores influence community structure. Impoverished faunas in scrub forests on white-sand soils show how plant defenses profoundly influence some animal communities. The toxin load and physical toughness of white-sand vegetation exclude most animal species, and sharply limit the abundance of others. Moreover, the presence or absence of critical mutualists can tilt the competitive relationships within communities. Keystone mutualists, in particular, may play pivotal roles in community organization.

The study of plant and animal relationships is rapidly becoming an integral part of population and community ecology. Not long ago, ecologists assumed that competition determined the distribution and abundance of plant species, and that predator control of herbivorous prey explained why the Earth is covered with uneaten plants. These are now insufficient explanations of either plant or animal abundance. Mandibles and teeth have a great deal to do with plant distribution and abundance, and thorns, lignins, and terpenes have much to offer in explaining why some animals are common while others are rare.

STUDY QUESTIONS

1. What distinguishes the effects of herbivory from the effects of predation?
2. Marine ecologist Jane Lubchenco (1986) found that snails in the intertidal zone feed on and suppress rapidly growing annual algae, thereby allowing perennial seaweeds to establish themselves. What parallels and contrasts can you make between this marine community and terrestrial communities?
3. How does the time scale of ecological interactions influence the interpretation of ecological studies?
4. The greater palatability and apparency of annual cucumbers growing in monocultures had little effect on their productivity of fruits, even though such monocultures attracted far more disease-carrying beetles than cucumbers growing in polycultures. How might a perennial life history change these results?
5. Is the wildebeest a keystone mutualist?
6. The World Wildlife Fund and Brazilian Government negotiate with Brazilian ranchers to leaves tracts of forest of different sizes as land is cleared for cattle raising. Assuming that refuges of different size lose different species to local extinction, how might this policy be used to test the keystone mutualist concept?
7. Should ecological relationships between animals and plants have a greater effect in determining community composition in forests on rich soils or in forests on poor soils?
8. Patrice Morrow (1971) notes that each year insects eat 10% of the foliage of African, American, Asian, and European forests, but that insects annually consume 20–50% of the foliage of Australian *Eucalyptus* forests. Why might this be so? What effects on the abundances of insect herbivores might be expected? On birds that eat insects? On *Eucalyptus* defenses?

SUGGESTED READING

Recent anthologies that offer insightful essays on the place of plant and animal relationships in ecology include Peter Price et al.'s (1984), *A New Ecology*; Douglas Boucher's

(1985), *The Biology of Mutualism: Ecology and Evolution:* and Jared Diamond and Ted Case's (1986), *Community Ecology.* Integrated discussions of an entire ecosystem may be found in A. R. E. Sinclair and M. Norton-Griffiths' (1979) *The Serengeti: Dynamics of an Ecosystem.*

Appendix I

Elementary Evolutionary Genetics

Population genetics evaluates allele frequency change without regard to phenotypes. Quantitative genetics evaluates phenotypic change without regard to the identity of alleles. Quantitative genetic theory is entirely derivable from population genetic theory (Falconer, 1981), but the practical applications of the two approaches differ.

Population genetics uses algebra to predict allele frequency change under idealized assumptions. Quantitative genetics predicts and measures response to selection of polygenic traits.

MENDELIAN TRAITS

The **theory of population genetics** predicts allele frequencies under different environmental circumstances. It is a genotype explicit theory because it concerns allele frequencies without reference to phenotypes. The beauty of population genetics is that it provides precise estimates of allele frequency change with a minimum of confounding assumptions. The tool for these estimates is to detect departure from the **Hardy–Weinberg equilibrium.**

The Hardy-Weinberg law states that the heredity process, by itself, does not change allele frequencies. If there are two alleles for one locus, A and a, with frequencies of p and q where $p + q = 1$, then the frequencies of the three possible genotypes in a population are

$$(p + q)^2 = p^2 + 2pq + q^2$$

$$A \quad a \qquad AA \quad Aa \quad aa$$

As long as assumptions of the Hardy–Weinberg law are met, these frequencies do not change from one generation to another. For instance, if the initial frequency of A in the population is .2 and the frequency of a is .8, the same frequencies will exist after 1, 1,000, or 1 million generations. Evolution does not occur.

A Hardy–Weinberg equilibrium holds if there is (1) no mutation, (2) no migration, (3) no selection, and (4) random breeding in a large population. If mutation, migration, or selection influences one allele (A or a above) more than the other, gene frequencies change and evolution occurs. Gene frequencies can also change if mating is not random, or if random pairing in small populations happens by chance to sample one allele more than the other.

The utility of this law is its power to specify allele frequency change when its assumptions are broken. Suppose that the genotypes AA, Aa, and aa have viabilities of .75, .75, and .68, respectively. These different **absolute fitnesses** of the genotypes show that selection is occurring. The easiest means of illustrating natural selection is with **relative fitnesses,** which can be calculated as $.75/.75 = 1$, $.75/.75 = 1$, and $.68/.75 = .9$. The genotype aa is only 90% as "fit" as AA or Aa.

This example shows, first, how gene frequencies can be altered by selection when A is dominant over a. A **coefficient of selection** can be calculated as $s = 1 - w_{aa}$, where w_{aa} is the relative fitness of aa. In short, $s = 1 - .9 = .1$. Presuming a constant magnitude of selection against genotype aa, with each generation the genotype frequencies of the population can be calculated as $p^2 + 2pq + q^2(1 - s)$. The change in $q = -sq^2(1 - q) / (1 - sq^2)$ (see Fig. 5-2).

This example illustrates selection on one locus. The Hardy–Weinberg principle can be expanded to predict genotype frequencies for alleles at any number of independent loci, with or without natural selection. The equation can similarly accommodate terms for mutation, migration, and genetic drift due to sampling errors or assortative mating.

POLYGENIC TRAITS

Polygenic traits are those influenced by allele differences at several or many loci. Phenotypic variance of a trait (Vp) can therefore be broken down into components:

$$V_P = V_A + V_D + V_I + V_E$$

where

V_P = total variance of the phenotype
V_A = variance of the phenotype due to additive gene differences
V_D = variance of the phenotype due to dominance effects
V_I = variance of the phenotype due to interactions between loci
V_E = variance of the phenotype due to environmental effects

Note the key point here is that variation in the phenotype can be due to any or all of three genetic sources of variance (or V_G), or to variation in the environment. If variation in gene differences is small, or environmental variation is large, gene influences on phenotype may be negligible. Some quantitative geneticists substitute a term for gene/environment interaction ($V_{G \times E}$) for V_E.

Heritability (h^2) is the extent to which a population can respond to selection on any given trait. It may be defined as

$$h^2 = R/S$$

where R is the response to selection, or the difference in means between a parent population and the offspring of a subset of that parent population allowed to breed, and S is the selection differential, or the difference in means between the parent population and the subset allowed to breed. Finally, one may calculate the intensity of selection (i) as

$$i = R/sh^2$$

where s is the standard deviation of the phenotype of the trait.

Heritability is also defined as the ratio of some measure of genetic variance over the total phenotypic variance in a trait:

$$\text{heritability in the broad sense} = h_b^2 = V_G/V_P$$

where $V_G = V_A + V_D + V_I$.

$$\text{heritability in the narrow sense} = h_n^2 = V_A/V_P$$

h_n^2 may be estimated from the slope of the linear regression of measures of offspring characteristics against similar measures of parental characteristics.

Appendix II

Measures of Growth and Feeding Efficiency Commonly Used in Comparisons of Herbivore Use of Plant Foods

AD   Approximate digestibility = % digestibility = [weight of food ingested (l in mg) − weight of feces (F in mg)]/weight of food ingested (l in mg) × 100

A   Absorption in the gut = weight of food ingested − weight of feces = $l − F$

B   Biomass increase = weight gain = weight (mg)/mean weight (mean mg)

ECD   Efficiency of conversion of digested biomass = % conversion efficiency − $(B/l − F) \times 100$

ECI   Overall efficiency of conversion = (AD × ECD) × 100

RCR   Relative consumption rate = food ingested (mg)/mean larval weight (mg)/day

RGR   Relative growth rate = RCR × AD × ECD = RCR × ECI

Sources: Erickson and Feeny (1974), Slansky and Feeny (1977), Van Soest (1982), and others.

Literature Cited

Addicott, J. F. (1981). Stability properties of two-species models of mutualism: Simulation studies. *Oecologia* **49:** 42–49.

Addicott, J. F. (1986). Variation in the costs and benefits of mutualism: The interaction between yuccas and yucca moths. *Oecologia* **70:** 486–494.

Agami, M., and Y. Waisel (1986). The role of mallard ducks *(Anas platyrhynchos)* in distribution and germination of seeds of the submerged hydrophyte *Najas marina* L. *Oecologia* **68:** 473–475.

Ågren, J., T. Elmqvist, and A. Tunlid (1986). Pollination by deceit, floral sex ratios and seed set in dioecious *Rubus chamaemorus* L. *Oecologia* **70:** 332–338.

Aker, C. L. (1982). Spatial and temporal dispersion patterns of pollinators and their relationship to the flowering strategy of *Yucca whipplei* (Agavaceae). *Oecologia* **54:** 243–252.

Alvarez, L. W., W. Alvarez, F. Asaro, and H. V. Michel (1980). Extraterrestrial causes for the Cretaceous-Tertiary extinction. *Science* **208:** 1095–1108.

Anderson, A. B. (1981). White-sand vegetation of Brazilian Amazonia. *Biotropica* **13:** 199–210.

Anderson, E. (1984). Who's who in the Pleistocene: A mammalian bestiary. Pages 40–89 *in* P. S. Martin and R. G. Klein, Editors. Quaternary Extinctions: A Prehistoric Revolution. University of Arizona Press, Tucson.

Argenzio, R. A. (1984). Digestion, absorption, and metabolism. Pages 262–277 *in* M. J. Swenson, Editor. Duke's Physiology of Domestic Animals. Tenth Edition. Cornell University Press, Ithaca, New York.

Bach, C. E. (1980). Effects of plant density and diversity on the population dynamics of a specialist herbivore, the striped cucumber beetle, *Acalymma vittata* (Fab.). *Ecology* **61:** 1515–1530.

Baker, H. G., and I. Baker (1983). Floral nectar sugar constituents in relation to pollinator type. Pages 117–141 *in* C. E. Jones and R. J. Little, Editors. Handbook of Experimental Pollination Biology. Van Nostrand Reinhold Company, New York.

Bakker, R. T. (1983). The deer flies, the wolf pursues: Incongruities in predator–prey coevolution. Pages 350–382 *in* D. Futuyma and M. Slatkin, Editors. Coevolution. Sinauer Associates, Sunderland, Massachusetts.

Barth, F. (1985). Insects and Flowers. Princeton University Press, Princeton.

Bartholomew, G. A., and T. M. Casey (1978). Oxygen consumption of moths during rest, pre-flight warm-up, and flight in relation to body size and wing morphology. *Journal of Experimental Biology* **76:** 11–25.

Bawa, K. (1980a). The evolution of dioecy in flowering plants. *Annual Review of Ecology and Systematics* **11:** 15–39.

Bawa, K. S. (1980b). Mimicry of male by female flowers and intrasexual competition for pollinators in *Jacaratia dolichaula* (D. Smith) Woodson (Caricaceae). *Evolution* **34:** 467–474.

Bawa, K. S. (1983). Patterns of flowering in tropical plants. Pages 394–410 *in* C. E. Jones and R. J. Little, Editors. Handbook of Experimental Population Biology. Van Nostrand Reinhold, New York.

Beattie, A. J. (1985). The Evolutionary Ecology of Ant-Plant Mutualisms. Cambridge University Press, Cambridge.

Beck, C., Editor (1976). Origin and Early Evolution of Angiosperms. Columbia University Press, New York.

Bell, R. H. V. (1970). The use of the herb layer by grazing ungulates in the Serengeti. Pages 111–124 *in* A. Watson, Editor. Animal Populations in Relation to their Food Resources. Blackwell Scientific Publications, Oxford.

Bell, W. J., and R. T. Carde, Editors (1984). Chemical Ecology of Insects. Chapman and Hall, London.

Belt, T. (1874). The Naturalist in Nicaragua. J. Murray, London. (Reprinted by the University of Chicago Press, 1985.)

Benson, W. W. (1982). Alternative models for infrageneric diversification in the humid tropics: Tests with passion vine butterflies. Pages 608–640 *in* G. T. Prance, Editor. Biological Diversification in the Tropics. Columbia University Press, New York.

Bentley, B. (1976). Plants bearing extrafloral nectaries and the associated ant community: Interhabitat differences in the reduction of herbivore damage. *Ecology* **54:** 815–820.

Bentley, B. and T. Elias, Editors (1983). The Biology of Nectaries. Columbia University Press, New York.

Berg, R. Y. (1969). Adaptation and evolution in *Dicentra* (Fumariaceae) with special reference to seed, fruit, and dispersal mechanism. *Nytt Magasin for Botanikk* **16:** 49–75.

Berenbaum, M. (1981). Patterns of furanocoumarin distribution and insect herbivory in the Umbelliferae: Plant chemistry and community structure. *Ecology* **63:** 1254–1266.

Berenbaum, M. (1983). Coumarins and caterpillars: A case for coevolution. *Evolution* **37:** 163–179.

Bernays, E. A. (1981). Plant tannins and insect herbivores: An appraisal. *Ecological Entomology* **6:** 353–360.

Bertin, R. I. (1985). Nonrandom fruit production in *Campsis radicans:* Between-year consistency and effects of prior pollination. *American Naturalist* **126:** 750–759.

Bierzychudek, P. (1982). The demography of Jack-in-the-pulpit, a forest perennial that changes sex. *Ecological Monographs* **52:** 335–351.

Bierzychudek, P. (1987). Pollinators increase the cost of sex by avoiding female flowers. *Ecology* **68:** 444–447.

Bock, W. J., R. P. Balda, and S. B. Vander Wall (1973). Morphology of the sublingual pouch and tongue musculature in Clark's nutcracker. *Auk* **90:** 491–519.

Bonaccorso, F. J. (1979). Foraging and reproductive biology in a Panamanian bat community. *Bulletin of the Florida State Museum: Biological Sciences* **24:** 360–408.

Borror, D., D. M. DeLong, and C. A. Triplehorn (1981). An Introduction to the Study of Insects, Fifth Edition. Saunders, New York.

Boucher, D. H., Editor (1985). The Biology of Mutualism: Ecology and Evolution. Oxford University Press, New York.

Bourlière, F. (1975). Mammals, small and large: The ecological implications of size.

Pages 1–8 *in* F. B. Golley, K. Petrusewicz, and L. Ryskowski, Editors. Small Mammals: Their Productivity and Population Dynamics. Cambridge University Press, London.

Brandon, R. M., and R. M. Burian (1984). Genes, Organisms, Populations, The MIT Press, Cambridge.

Brattsten, L. (1979a). Biochemical defense mechanisms in herbivores against plant allelochemicals. Pages 200–270 *in* G. A. Rosenthal and D. H. Janzen, Editors. Herbivores: Their Interaction with Secondary Plant Metabolites. Academic Press, New York.

Brattsten, L. (1979b). Ecological significance of mixed-function oxidations. *Drug Metabolism Reviews* **10:** 35–58.

Brattsten, L. (1983). Cytochrome *P*-450 involvement in the interactions between plant terpenes and insect herbivores. Pages 173–195 *in* P. A. Hedin, Editor. Plant Resistance to Insects. American Chemical Society.

Brattsten, L. B., C. W. Holyoke, Jr., J. R. Leeper, and K. F. Raffa (1986). Insecticide resistance: Challenge to pest management and basic research. *Science* **231:** 1255–1260.

Brower, L. P. (1969). Ecological chemistry. *Scientific American* **220:** 22–29.

Brown, J. H., O. J. Reichman, and D. W. Davidson (1979). Granivory in desert ecosystems. *Annual Review of Ecology and Systematics* **10:** 201–227.

Bryant J. P. (1981). Phytochemical deterrence of snowshoe hare browsing by adventitious shoots of four Alaskan trees. *Science* **213:** 889–890.

Bryant, J. P., and P. J. Kuropat (1980). Selection of winter forage by subarctic browsing vertebrates: The role of plant chemistry. *Annual Review of Ecology and Systematics* **11:** 261–285.

Bryant J. P., F. S. Chapin III, and D. R. Klein (1983). Carbon/nutrient balance of boreal plants in relation to vertebrate herbivory. *Oikos* **40:** 357–368.

Bryant, J. P., F. S. Chapin III, P. Reichardt, and T. Clausen (1985). Adaptation to resource availability as a determinant of chemical defense strategies in woody plants. Pages 219–237 *in* G. A. Cooper-Driver, T. Swain, and E. E. Conn, Editors. Chemically Mediated Interactions between Plants and other Organisms. Plenum, New York.

Buchmann, S. L. (1983). Buzz pollination in angiosperms. Pages 73–116 *in* C. E. Jones and R. J. Little, Editors. Handbook of Experimental Pollination Biology. Van Nostrand Reinhold Company, New York.

Buchner P. (1965). Endosymbiosis in Insects. Wiley, New York.

Burbidge, A. H., and R. J. Whelan (1982). Seed dispersal in a cycad, *Macrozamia riedlei. Australian Journal of Ecology* **7:** 63–67.

Burnett, W. C.. Jr., S. B. Jones, Jr., and T. J. Mabry (1978). The role of sesquiterpene lactones in plant–animal coevolution. Pages 233–257 *in* J. B. Harborne, Editor. Biochemical Aspects of Plant and Animal Coevolution. Academic Press, London.

Campbell, D. R. (1985). Pollen and gene dispersal: The influences of competition for pollination. *Evolution* **39:** 418–431.

Campbell, D. R., and A. F. Motten (1985). The mechanism of competition for pollination between two forest herbs. *Ecology* **66:** 554–564.

Carpenter, F. L., and R. E. MacMillen (1976). Threshold model of feeding territoriality and test with a Hawaiian honeycreeper. *Science* **194:** 639–642.

Carlquist, S. (1974). Island Biology. Columbia University Press, New York.

Cech, T. R. (1985). Self-splicing RNA: Implications for evolution. *International Review of Cytology* **93:** 3–22.

Claridge, M. F., and J. Den Hollander (1980). The "biotypes" of the rice brown planthopper, *Nilaparvata lugens*. *Entomologia Experimentalis et Applicata* **27**: 23–30.

Claridge, M. F., and J. Den Hollander (1982a). Virulence to rice cultivars and selection for virulence in populations of the brown planthopper *Nilaparvata lugens*. *Entomologia Experimentalis et Applicata* **27**: 213–221.

Claridge, M. F., J. Den Hollander, and I. Furet (1982b). Adaptations of brown planthopper *(Nilaparvata lugens)* populations to rice varieties in Sri Lanka. *Entomologia Experimentalis et Applicata* **32**: 222–226.

Coley, P. D. (1983). Herbivory and defensive characteristics of tree species in a lowland tropical forest. *Ecological Monographs* **53**: 209–233.

Coley, P. D. J. P. Bryant, and F. S. Chapin III. (1985). Resource availability and plant antiherbivore defense. *Science* **230**: 895–899.

Conn, E. E. (1979). Cyanide and cyanogenic glycosides. Pages 387–412 *in* G. A. Rosenthal and D. H. Janzen, Editors. Herbivores: Their Interaction with Secondary Plant Metabolites. Academic Press, New York.

Connell, J. H. (1971). On the role of natural enemies in preventing competitive exclusion in some marine animals and in rain forest. Pages 298–312 *in* P. J. Den Boer and G. R. Gradwell, Editors. Dynamics of Populations. PUDOC, Wageningen, Netherlands.

Cook, R. (1980). The biology of seeds in the soil. Pages 107–130 *in* O. T. Solbrig, Editor. Demography and Evolution in Plant Populations. University of California Press, Berkeley.

Cooper, S. M., and N. Owen-Smith (1986). Effects of plant spinescence on large mammalian herbivores. *Oecologia* **68**: 446–455.

Cox, P. A. (1983). Extinction of the Hawaiian avifauna resulted in a change of pollinators for the ieie, *Freycinetia arborea. Oikos* **41**: 195–199.

Crawley, M. J. (1983). Herbivory. University of California Press, Berkeley.

Creighton, W. S. (1950). The ants of North America. *Bulletin of the Museum of Comparative Zoology, Harvard University* **104**: 1–585.

Crepet, W. L. (1983). The role of insect pollination in the evolution of the angiosperms. Pages 29–50 *in* L. Real, Editor. Pollination Biology. Academic Press, New York.

Cruden, R. W. (1976). Intraspecific variation in pollen-ovule ratios and nectar secretion—preliminary evidence of ecotypic adaptation. *Annals of the Missouri Botanical Garden* **63**: 277–289.

Cruden, R. W. (1977). Pollen-ovule ratios: A conservative indicator of breeding systems in flowering plants. *Evolution* **31**: 32–46.

Culver, D. C., and A. J. Beattie (1978). Myrmecochory in *Viola:* Dynamics of seed–ant interactions in some West Virginia species. *Journal of Ecology* **66**: 53–72.

Culver, D. C., and A. J. Beattie (1980). The fate of *Viola* seeds dispersed by ants. *American Journal of Botany* **67**: 710–714.

Darwin, C. (1859). On the Origin of Species. John Murray, London.

Davidson, D. W., and S. R. Morton (1981). Competition for dispersal in ant-dispersed plants. *Science* **213**: 1259–1261.

Davis, M. B. (1976). Pleistocene biogeography of temperate deciduous forests. *Geoscience and Man* **13**: 13–26.

Davis, M. B. (1986). Climatic instability, time lags, and community disequilibrium. Pages 269–284 *in* J. Diamond and T. J. Case, Editors. Community Ecology. Harper and Row, New York.

Denno, R. F., and M. S. McClure, Editors (1983). Variable Plants and Herbivores in Natural and Managed Systems. Academic Press, New York.

Dethier, V. G. (1954). Evolution of feeding preferences in phytophagous insects. *Evolution* **8**: 33–54.

Diamond, J., and T. J. Case, Editors (1986). Community Ecology. Harper and Row, New York.

Dirzo, R., and J. L. Harper (1982a). Experimental studies on slug–plant interactions. III. Differences in the acceptability of individual plants of *Trifolium repens* to slugs and snails. *Journal of Ecology* **70**: 101–117.

Dirzo, R., and J. L. Harper (1982b). Experimental studies on slug–plant interactions. IV. The performance of cyanogenic and acyanogenic morphs of *Trifolium repens* in the field. *Journal of Ecology* **70**: 119–138.

Docters van Leeuwen, W. M. (1954). On the biology of some Javanese Loranthaceae and the role birds play in their life-histories. *Beaufortia* **41**: 105–206.

Doyle, J. A., and L. J. Hickey (1976). Pollen and leaves from the Mid-Cretaceous Potomac group and their bearing on early angiopsperm evolution. Pages 139–206 *in* C. Beck, Editor. Origin and Early Evolution of Angiosperms. Columbia University Press, New York.

Dressler, R. L. (1981). The Orchids: Natural History and Classification. Harvard University Press, Cambridge.

Dyson, F. J. (1982). A model for the origin of life. *Journal of Molecular Evolution* **18**: 344–350.

Edmunds, G. F., Jr., and D. N. Alstad (1978). Coevolution in insect herbivores and conifers. *Science* **199**: 941–945.

Ehrlich, P., and P. H. Raven (1964). Butterflies and plants: A study in coevolution. *Evolution* **18**: 586–608.

Eigen, M., and P. Schuster (1982). Stages of emerging life—five principles of early organization. *Journal of Molecular Evolution* **19**: 47–61.

Ellstrand, N. C. (1984). Multiple paternity within the fruits of the wild radish, *Raphanus sativus*. *American Naturalist* **123**: 819–828.

Elton, C., and M. Nicholson (1942). The ten-year cycles in numbers of the lynx in Canada. *Journal of Animal Ecology* **11**: 215–244.

Erickson, J. M., and P. Feeny (1974). Sinigrin: A chemical barrier to the black swallowtail butterfly, *Papilio polyxenes*. *Ecology* **55**: 103–111.

Falconer, D. S. (1981). Introduction to Quantitative Genetics. Second Edition. Longman Press, London.

Feeny, P. (1970). Seasonal changes in oak leaf tannins and nutrients as a cause of spring feeding by winter moth caterpillars. *Ecology* **51**: 565–581.

Feeny, P. (1976). Plant apparency and chemical defense. *Recent Advances in Phytochemistry* **10**: 1–40.

Feinsinger, P., L. A. Swarm, and J. A. Wolfe (1985). Nectar-feeding birds on Trinidad and Tobago: Comparison of diverse and depauperate guilds. *Ecological Monographs* **55**: 1–28.

Flor, H. H. (1956). The complementary genic systems in flax and flax rust. *Advances in Genetics* **8**: 29–54.

Foster, R. B. (1982a). The seasonal rhythm of fruit fall on Barro Colorado Island. Pages 151–172, *in* E. G. Leigh, A. S. Rand, and D. S. Windsor, Editors. The Ecology of a Tropical Forest: Seasonal Rhythms and Long-term Changes. Smithsonian Press, Washington, D.C.

Foster, R. B. (1982b). Famine on Barro Colorado Island. Pages 201–212, *in* E. G.

Leigh, A. S. Rand, and D. S. Windsor, Editors. The Ecology of a Tropical Forest: Seasonal Rhythms and Long-term Changes. Smithsonian Press, Washington, D.C.

Fox, L. R. (1981). Defense and dynamics in plant–herbivore systems. *American Zoologist* **21**: 853–864.

Fox, L. R., and P. A. Morrow. (1981). Specialization: Species property or local phenomenon? *Science* **211**: 887–893.

Fox, S. W., and K. Dose (1972). Molecular Evolution and the Origin of Life. W. H. Freeman and Company, San Francisco.

Fraenkel, G. S. (1959). The *raison d'etre* of secondary plant substances. *Science* **129**: 473–486.

Freeland, W. J., and D. H. Janzen (1974). Strategies in herbivory by mammals: The role of plant secondary compounds. *American Naturalist* **108**: 269–289.

Futuyma, D. (1976). Food plant specialization and environmental predictability in Lepidoptera. *American Naturalist* **110**: 285–292.

Futuyma, D. D. (1986). Evolutionary Biology. Second Edition. Sinauer Associates, Sunderland, Massachusetts.

Futuyma, D., and S. C. Peterson (1985). Genetic variation in the use of resources by insects. *Annual Review of Entomology* **30**: 217–238.

Futuyma, D., and M. Slatkin, Editors (1983). Coevolution. Sinauer Associates, Sunderland, Massachusetts.

Gaines, S. D., and J. Lubchenco (1982). A unified approach to marine plant–herbivore interactions. II. Biogeography. *Annual Review of Ecology and Systematics* **13**: 111–138.

Galil, J. (1977). Fig biology. *Endeavour* : 52–56.

Gallun, R. L., and G. S. Khush (1980). Genetic factors affecting expression and stability of resistance. Pages 63–85 *in* F. G. Maxwell and P. R. Jennings, Editors. Breeding Plants Resistant to Insects. John Wiley and Sons, New York.

Gause, G. F., and A. A. Witt (1935). Behavior of mixed populations and the problem of natural selection. *American Naturalist* **69**: 596–609.

Gautier-Hion, A., J. M. Duplantier, R. Quris, F. Feer, C. Sourd, J. P. Decoux, G. Dubost, L. Emmons, C. Erard, P. Hecketsweiler, A. Moungazi, C. Roussilhon, and J. M. Thiollay (1985). Fruit characters as a basis of fruit choice and seed dispersal in a tropical forest vertebrate community. *Oecologia* **65**: 324–337.

Gilbert, L. E. (1980). Food web organization and the conservation of neotropical diversity. Pages 11–33 *in* M. E. Soule and B. A. Wilcox, Editors. Conservation Biology. Sinauer Associates, Sunderland, Massachusetts.

Gilpin, M. (1973). Do hares eat lynx? *American Naturalist* **107**: 727–730.

Gould, F. (1983). Genetics of plant-herbivory systems: Interactions between applied and basic study. Pages 599–654 *in* R. F. Denno and M. S. McClure, Editors. Variable Plants and Herbivores in Natural and Managed Systems. Academic Press, New York.

Grant, P. R., J. N. M. Smith, B. R. Grant, I. J. Abbott, and L. K. Abbott (1975). Finch numbers, owl predation and plant dispersal on Isla Daphne Major, Galapagos. *Oecologia* **19**: 239–257.

Grant, V., and K. Grant (1965). Flower Pollination in the Phlox Family. Columbia University Press, New York.

Graham, R. W. (1986). Response of mammalian communities to environmental changes during the late Quarternary. Pages 300–313 *in* J. Diamond and T. J. Case, Editors. Community Ecology. Harper and Row, New York.

Grinnell, J. (1917). The niche relationships of the California thrasher. *Auk* **34**: 427–433.

Guthrie, R. D. (1984). Mosaics, allelochemics and nutrients: An ecological theory of late Pleistocene megafaunal extinctions. Pages 259–298 *in* P. S. Martin and R. G. Klein, Editors. Quaternary Extinctions: A Prehistoric Revolution. University of Arizona Press, Tucson.

Handel, S. N., S. B. Fisch, and G. E. Schatz (1981). Ants disperse a majority of herbs in a mesic forest community in New York State. *Bulletin of the Torrey Botanical Club* **108**: 430–437.

Harborne, J. B. (1977). Introduction to Ecological Biochemistry. Academic Press, New York.

Hartl, D. (1981). A Primer in Population Genetics. Sinauer Associates, Sunderland, Massachusetts.

Haukioja, E. (1980). On the role of plant defenses in the fluctuation of herbivore populations. *Oikos* **35**: 202–213.

Haukioja, E., P. Niemela, and S. Siren (1985a). Foliage phenols and nitrogen in relation to growth, insect damage, and ability to recover after defoliation, in the mountain birch *Betula pubescens ssp. tortuosa*. *Oecologia* **65**: 214–222.

Haukioja, E., J. Suomala, and S. Neuvonen. (1985b). Long-term inducible resistance in birch foliage: Triggering cues and efficiency on a defoliator. *Oecologia* **65**: 363–369.

Heinrich, B. (1976). Foraging specializations of individual bumblebees. *Ecological Monographs* **46**: 105–128.

Heinrich, B. (1979a). "Majoring" and "minoring" by foraging bumblebees, *Bombus vagans:* An experimental analysis. *Ecology* **60**: 245–255.

Heinrich, B. (1979b). Bumblebee Economics. Harvard University Press, Cambridge, Massachusetts.

Heinrich, B., and P. Raven (1972). Energetics of pollination. *Science* **176**: 597–602.

Heithaus, E. R., T. H. Fleming, and P. A. Opler (1975). Foraging patterns and resource utilization in seven species of bats in a seasonal tropical forest. *Ecology* **56**: 841–854.

Hendrix, S. D. (1984). Reactions of *Heracleum lanatum* to floral herbivory by *Depressaria pastinacella*. *Ecology* **65**: 191–197.

Herrera, C. M. (1981). Are tropical fruits more rewarding to dispersers than temperate ones? *American Naturalist* **118**: 896–907.

Herrera, C. M. (1984a). Seed dispersal and fitness determinants in wild rose: Combined effects of hawthorn, birds, mice, and browsing ungulates. *Oecologia* **63**: 386–393.

Herrera, C. M. (1984b). Adaptation to frugivory of Mediterranean avian seed dispersers. *Ecology* **65**: 609–617.

Herrera C. M. (1984c). A study of avian frugivores, bird-dispersed plants, and their interaction in Mediterranean scrublands. *Ecological Monographs* **54**: 1–23.

Herrera, C. M. (1985). Determinants of plant–animal coevolution: The case of mutualistic vertebrate seed dispersal systems. *Oikos* **44**: 132–141.

Hladik, C. M. (1967). Surface relative du tractus digestif de quelques primates. Morphologie des villosites intestinales et correlations avec la regime alimentaire. *Mammalia* **31**: 120–147.

Horvitz, C. C., and D. W. Schemske (1984). Effects of ants and an ant-tended herbivore on seed production of a neotropical herb. *Ecology* **65**: 1369–1378.

Howard, J. J., J. Cazin, Jr., and D. F. Wiemer (1988). Toxicity of terpenoid deterrents to the leafcutting ant *Atta cephalotes* and its mutualistic fungus. *Journal of Chemical Ecology* **14**: 59–68.

Howe, H. F. (1977). Bird activity and seed dispersal of a tropical wet forest tree. *Ecology* **58**: 539–550.

Howe, H. F. (1980). Monkey dispersal and waste of a neotropical fruit. *Ecology* **61**: 944–959.

Howe, H. F. (1983). Annual variation in a neotropical seed-dispersal system. Pages 211–227 *in* S. L. Sutton, T. C. Whitmore, and A. C. Chadwick, Editors. Tropical Rain Forest: Ecology and Management. Blackwell Scientific, Oxford.

Howe, H. F. (1984a). Constraints on the evolution of mutualisms. *American Naturalist* **123**: 764–777.

Howe, H. F. (1984b). Implications of seed dispersal by animals for the management of tropical reserves. *Biological Conservation* **30**: 261–281.

Howe, H. F. (1985). Gomphothere fruits: A critique. *American Naturalist* **125**: 853–865.

Howe, H. F. (1986). Seed dispersal by fruit-eating birds and mammals. Pages 123–190 *in* D. R. Murray, Editor. Seed Dispersal. Academic Press, Sydney.

Howe, H. F., E. W. Schupp, and L. C. Westley (1985). Early consequences of seed dispersal for a neotropical tree *(Virola surinamensis)*. *Ecology* **66**: 781–791.

Howe, H. F., and J. Smallwood (1982). Ecology of seed dispersal. *Annual Review of Ecology and Systematics* **13**: 201–228.

Howe, H. F., and G. A. Vande Kerckhove (1979). Fecundity and seed dispersal of a tropical tree. *Ecology* **60**: 180–189.

Howe, H. F., and G. A. Vande Kerckhove (1980). Nutmeg dispersal by tropical birds. *Science* **210**: 925–927.

Howe, H. F., and G. A. Vande Kerckhove (1981). Removal of wild nutmeg *(Virola surinamensis)* crops by birds. *Ecology* **62**: 1093–1106.

Howe, H. F., and L. C. Westley (1986). Ecology of pollination and seed dispersal. Pages 185–215 *in* M. J. Crawley, Editor. Plant Ecology. Blackwell Scientific Publications, London.

Hubbell, S. P., and D. F. Wiemer (1983). Host plant selection by an attine ant. Pages 133–154 *in* P. Jaisson, Editor. Social Insects in the Tropics, Volume 2. University of Paris, Paris.

Hubbell, S. P., J. J. Howard, and D. F. Wiemer (1984). Chemical leaf repellency to an attine ant: Seasonal distribution among potential host plant species. *Ecology* **65**: 1067–1076.

Hucker, H. B. (1970). Species differences in drug metabolism. *Annual Review of Pharmacology* **10**: 99–118.

Huheey, J. E. (1984). Warning coloration and mimicry. Pages 257–300 *in* W. J. Bell and R. T. Carde, Editors. Chemical Ecology of Insects. Chapman and Hall, New York.

Hutchinson, G. E. (1959). Homage to Santa Rosalia *or* Why are there so many kinds of animals? *American Naturalist* **93**: 145–159.

Inouye, D. W. (1977). Species structure of bumblebee communities in North America and Europe. Pages 35–40 *in* W. J. Mattson, Editor. The Role of Arthropods in Forest Ecosystems. Springer-Verlag, New York.

Jaenike, J. (1985). Genetic and environmental determinants of food preference in *Drosophila tripunctata*. *Evolution* **39**: 362–369.

Janson, C. (1983). Adaptation of fruit morphology to dispersal agents in a neotropical forest. *Science* **219**: 187–189.

Janzen, D. H. (1966). Coevolution of mutualism between ants and acacias in Central America. *Evolution* **20**: 249–275.

Janzen, D. H. (1970). Herbivores and the number of tree species in tropical forests. *American Naturalist* **104**: 501–528.

Janzen, D. H. (1972). Protection of *Barteria* (Passifloraceae) by *Pachysima* ants (Pseudomyrmecinae) in a Nigerian rain forest. *Ecology* **53**: 885–892.

Janzen, D. H. (1974). Tropical blackwater rivers, animals, and mast fruiting of the Dipterocarpaceae. *Biotropica* **6**: 69–113.

Janzen, D. H. (1978). Complications in interpreting the chemical defenses of trees against tropical arboreal plant-eating vertebrates. Pages 73–84 *in* G. G. Montgomery, Editor. The Ecology of Arboreal Folivores. Smithsonian Institution Press, Washington, D. C.

Janzen, D. H. (1980). When is it coevolution? *Evolution* **34**: 611–612.

Janzen, D. H. (1981). *Enterolobium cyclocarpum* seed passage rate and survival in horses, Costa Rican Pleistocene seed dispersal agents. *Ecology* **62**: 593–601.

Janzen, D. H. (1986). Chihuahuan desert nopaleras: Defaunated big mammal vegetation. *Annual Review of Ecology and Systematics* **17**: 595–636.

Janzen, D. H., and P. Martin (1982). Neotropical anachronisms: What the gomphotheres ate. *Science* **215**: 19–27.

Jarman, P. J., and A. R. E. Sinclair (1979). Feeding strategy and the pattern of resource-partioning in ungulates. Pages 130–163 *in* A. R. E. Sinclair and M. Norton-Griffiths, Editors. Serengeti: Dynamics of an Ecosystem. University of Chicago Press, Chicago

Johnson, L. K., and S. P. Hubbell (1974). Aggression and competition among stingless bees: Field studies. *Ecology* **55**: 120–127.

Jolivet, P. (1985). Un myrmecophyte hors de son pays d'origine: *Clerodendrum fallax* Lindley, 1844 (Verbenaceae) aux Iles du Cap Vert. *Bull. Soc. Linn. Lyon* **54**: 122–128.

Jones, C. E., and R. J. Little, Editors (1983). Handbook of Experimental Pollination Biology. Van Nostrand Rheinhold, New York.

Jones, C. G. (1984). Microorganisms as mediators of plant resource exploitation by insect herbivores. Page 53–99 *in* P. W. Price, C. N. Slobodchikoff, and W. S. Gaud, Editors. A New Ecology: Novel Approaches to Interactive Systems. John Wiley and Sons, New York.

Jones, D. A., R. J. Keymer, and W. M. Ellis (1978). Cyanogenesis in plants and animal feeding. Pages 21–34 *in* J. B. Harborne, Editor. Biochemical Aspects of Plant and Animal Coevolution. Academic Press, London.

Kareiva, P. (1983). Influence of vegetation texture on herbivore populations: Resource concentration and herbivore movement. Pages 259–289 *in* R. Denno and M. McClure, Editors. Variable Plants and Herbivores in Natural and Managed Systems. Academic Press, New York.

Keeler, K. H. (1985). Cost:benefit models of mutualism. Pages 100–127 *in* D. Boucher, Editor. The Biology of Mutualisms: Ecology and Evolution. Oxford University Press, Oxford.

Keith, L. B. (1983). Role of food in hare population cycles. *Oikos* **40**: 385–395.

Kevan, P. G. (1983). Floral colors through the insect eye: What they are and what they mean. Pages 3–30 *in* C. E. Jones and R. J. Little, Editors. Handbook of Experimental Pollination Biology. Van Nostrand Reinhold, New York.

Knerer, G., and C. E. Atwood (1973). Diprionid sawflies: Polymorphism and speciation. *Science* **179**: 1090–1099.

Knoll, A. H. (1986). Patterns of change in plant communities through geological time. Pages 126–141 *in* J. Diamond and T. G. Case, Editors. Community Ecology. Harper and Row, New York.

Koch, A. (1967). Insects and their endosymbionts. Pages 1–106 *in* S. M. Henry, Editor. Symbiosis II. Academic Press, New York.

Koptur, S. (1979). Facultative mutualism between weedy vetches bearing extrafloral nectaries and weedy ants in California. *American Journal of Botany* **66:** 1016–1020.

Krebs, C. J. (1985). Ecology: An Experimental Approach. Harper and Row. New York.

Krieger, R. I., P. P. Feeny, and C. F. Wilkinson (1971). Detoxification enzymes in the guts of caterpillars: An evolutionary answer to plant defenses? *Science* **172:** 579–581.

Kuhn, T. (1970). The Structure of Scientific Revolutions. Second Edition. University of Chicago Press, Chicago.

Kukor, J. J., and M. M. Martin (1986). The transformation of *Saperda calcarata* (Coleoptera: Cerambycidae) into a cellulose digester through the inclusion of fungal enzymes in its diet. *Oecologia* **71:** 138–141.

Kurten, B., and E. Anderson (1980). Pleistocene Mammals of North America. Columbia University Press, New York.

Lakatos, I., and A. Musgrave, Editors (1970). Criticism and the Growth of Knowledge. Cambridge University Press, Cambridge.

Law, R. (1985). Evolution in a mutualistic environment. Pages 145–171 *in* D. Boucher, Editor. The Biology of Mutualisms: Ecology and Evolution. Oxford University Press, Oxford.

Leigh, E. G., Jr., A. S. Rand, and D. Windsor, Editors (1982). The Ecology of a Tropical Forest: Seasonal Rhythms and Long-term Changes. Smithsonian Press, Washington, D.C.

Leppik, E. E. (1972). Origin and evolution of bilateral symmetry in flowers. *Evolutionary Biology* **5:** 49–86.

Lertzman, K. P., and C. L. Gass (1983). Alternative models of pollen transfer. Pages 474–490 *in* C. E. Jones and R. J. Little, Editors. Handbook of Experimental Pollination Biology. Van Nostrand Reinhold, New York.

Levene, H. (1953). Genetic equilibrium when more than one ecological niche is available. *American Naturalist* **87:** 311–313.

Levey, D. (1987). Seed size and fruit-handling techniques of avian frugivores. *American Naturalist* **129:** 471–485.

Levin, D. A., and H. W. Kerster (1968). Local gene dispersal in *Phlox pilosa. Evolution* **22:** 130–139.

Levin, D. A., and H. W. Kerster (1974). Gene flow in seed plants. *Evolutionary Biology* **7:** 139–220.

Levin, D. A., and A. C. Wilson (1976). Rates of evolution in seed plants: Net increase in diversity of chromosome numbers and species numbers through time. *Proceedings of the National Academy of Sciences, U.S.A.* **73:** 2086–2090.

Levin, S. A. (1983). Some approaches to the modelling of coevolutionary interactions. Pages 21–66 *in* M. H. Nitecki, Editor. Coevolution. University of Chicago Press, Chicago.

Levins, R. (1968). Evolution in Changing Environments. Princeton University Press, Princeton.

Lewontin, R. (1978). Adaptation. *Scientific American* **239:** 212–230.

Linhart, Y. B., and P. Feinsinger (1980). Plant–hummingbird interactions: Effects of island size and degree of specialization on pollination. *Journal of Ecology* **68:** 745–760.

Linsley, E. G., J. W. MacSwain, and P. H. Raven (1963). Comparative behavior of bees and Onagraceae. I. *Oenothera* bees of the Colorado Desert. II. *Oenothera* bees of the Great Basin. *University of California Publications in Entomology* **33:** 1–58.

Linsley, E. G., J. W. MacSwain, and P. H. Raven (1964). Comparative behavior of bees and Onagraceae. III. *Oenothera* bees of the Mojave Desert, California. *University of California Publications in Entomology* **33:** 59–98.

Lloyd, D. G., and K. S. Bawa (1984). Modification of the gender of seed plants in varying conditions. *Evolutionary Biology* **17:** 255–336.

Louda, S. M. (1982a). Distribution ecology: Variation in plant recruitment over a gradient in relation to insect seed predation. *Ecological Monographs* **52:** 25–41.

Louda, S. M. (1982b). Limitation of the recruitment of the shrub *Haplopappus squarrosus* (Asteraceae) by flower- and seed-feeding insects. *Journal of ecology* **70:** 43–53.

Lubchenco, J. (1986). Relative importance of competition and predation: Early colonization by seaweeds in New England. Pages 537–555 *in* J. Diamond and T. J. Case, Editors. Community Ecology. Harper & Row, New York.

Lubchenco, J., and S. D. Gaines (1981). A unified approach to marine plant–herbivore interactions. I. Populations and communities. *Annual Review of Ecology and Systematics* **12:** 405–437.

Mabry, T. J., and J. E. Gill (1979). Sesquiterpene lactones and other terpenoids. Pages 502–538 *in* G. A. Rosenthal and D. H. Janzen, Editors. Herbivores: Their Interactions with Secondary Plant Metabolites. Academic Press, New York.

MacArthur, R. H., and E. O. Wilson (1967). The Theory of Island Biogeography. Princeton University Press, Princeton.

Marks, P. L. (1974). The role of pin cherry (*Prunus pensylvanica* L.) in the maintenance of stability in northern hardwood ecosystems. *Ecological Monographs* **44:** 73–88.

Marquis, R. J. (1984). Leaf herbivores decrease fitness of a tropical plant. *Science* **226:** 537–539.

Martin, P. S. (1967). Prehistoric overkill. Pages 75–120 *in* P. S. Martin and H. E. Wright, Jr., Editors. Pleistocene Extinctions: A Search for a Cause. Yale University Press, New Haven.

Martin, P. S. (1983). Catastrophic extinctions and late Pleistocene blitzkrieg: Two radiocarbon tests. Pages 153–190 *in* M. H. Nitecki, Editor. Extinctions. University of Chicago Press, Chicago.

Martin, P. S., and R. G. Klein, Editors (1984). Quaternary Extinctions: A Prehistoric Revolution. University of Arizona Press, Tucson.

Martin, T. E. (1985). Resource selection by tropical frugivorous birds: Integrating multiple interactions. *Oecologia* **66:** 563–573.

Maynard Smith, J. (1978). Optimization theory in evolution. *Annual Review of Ecology and Systematics* **9:** 31–55.

Maxwell, F. G., and P. R. Jennings, Editors (1980). Breeding Plants Resistant to Insects. John Wiley and Sons, New York.

McBee, R. H. (1971). Significance of intestinal microflora in herbivory. *Annual Review of Ecology and Systematics* **2:** 165–176.

McKey, D. (1979). The distribution of secondary compounds within plants. Pages 56–

134 *in* G. A. Rosenthal and D. H. Janzen, Editors. Herbivores: Their Interaction with Secondary Plant Metabolites. Academic Press, New York.

McKey, D. (1984). Interaction of the ant-plant *Leonardoxa africana* (Caesalpiniaceae) with its obligate inhabitants in a rainforest in Cameroon. *Biotropica* **16:** 81–99.

McKey, D. B., P. G. Waterman, C. N. Mbi, J. S. Gartlan, and T. H. Struthsaker (1978). Phenolic content of vegetation in two African rain forests: Ecological implications. *Science* **202:** 61–64.

McKey, D. B., J. S. Gartlan, P. G. Waterman, and G. M. Choo (1981). Food selection by black colobus monkeys *(Colobus satanas)* in relation to plant chemistry. *Biological Journal of the Linnean Society* **16:** 115–146.

McNaughton, S. J. (1976). Serengeti migratory wildebeest: Facilitation of energy flow by grazing. *Science* **191:** 92–94.

McNaughton, S. J. (1978). Serengeti ungulates: Feeding selectivity influences the effectiveness of plant defense guilds. *Science* **199:** 806–807.

McNaughton, S. J. (1984). Grazing lawns: Animals in herds, plant form, and coevolution. *American Naturalist* **124:** 863–886.

McNaughton, S. J. (1985). Ecology of a grazing ecosystem: The Serengeti. *Ecological Monographs* **55:** 259–294.

McNaughton, S. J., and J. L. Tarrants (1983). Grass leaf silicification: Natural selection for an inducible defense against herbivores. *Proceedings of the National Academy of Sciences*, U.S.A. **80:** 790–791.

Milchunas, D. G. (1977). *In vivo-in vitro* relationships of Colorado mule deer forages. Master of Science Thesis, Colorado State University, Fort Collins, Colorado.

Miller, J. R., and K. L. Strickler (1984). Finding and accepting host plants. Pages 127–158 *in* W. J. Bell and R. T. Carde, Editors. Chemical Ecology of Insects. Chapman and Hall, New York.

Miller, S. L. (1953). A production of amino acids under possible primitive earth conditions. *Science* **117:** 528–529.

Milton, K. (1978). Behavioral adaptation to leaf-eating by the mantled howler monkey *(Alouatta palliata)*. Pages 535–550 *in* G. G. Montgomery, Editor. The Ecology of Arboreal Folivores. Smithsonian Institution Press, Washington, D. C.

Mitter, C. and D. R. Brooks (1983). Phylogenetic aspects of coevolution. Pages 65–98 *in* D. Futuyma and M. Slatkin, Editors. Coevolution. Sinauer Associates, Sunderland, Massachusetts.

Moermond, T. C., and J. S. Denslow (1983). Fruit choice in tropical frugivorous birds: Effects of fruit type and accessibility on selectivity. *Journal Animal Ecology* **52:** 407–421.

Moermond, T. C., and J. S. Denslow (1985). Neotropical avian frugivores: Patterns of behavior, morphology, and nutrition with consequences for fruit selection. Pages 865–897 *in* P. A. Buckley, M. S. Foster, E. S. Morton, R. S. Ridgely, and N. G. Smith, Editors. Neotropical Ornithology. A.O.U. Monographs.

Montagu, A. Editor (1984). Science and Creationism. Oxford University Press, New York.

Morrison, D. (1978). Foraging ecology and energetics of the frugivorous bat *Artibeus jamaicensis*. *Ecology* **59:** 716–723.

Morrow, P. A. (1977). The significance of phytophagous insects in the *Eucalyptus* forests of Australia. Pages 19–29 *in* W. J. Mattson, Editor. The Role of Arthropods in Forest Ecosystems. Springer-Verlag, New York.

Mulkey, S. S., A. P. Smith, and T. P. Young (1984). Predation by elephants on *Se-*

necio keniodendron (Compositae) in the alpine zone of Mount Kenya. *Biotropica* **16**: 246–248.

Murray, D., Editor (1986). Seed Dispersal. Academic Press, Sydney.

Norris, D. M., and M. Kogan (1980). Biochemical and morphological bases of resistance. Pages 23–62 *in* F. G. Maxwell and P. R. Jennings, Editors. Breeding Plants Resistant to Insects. John Wiley and Sons, New York.

Officer, C. B., and C. L. Drake (1985). Terminal Cretaceous environmental events. *Science* **227**: 1161–1167.

Oparin, A. I. (1936). The Origin of Life on Earth. Second Edition. Dover, New York.

Owen, D. F. (1980). How plants may benefit from the animals that eat them. *Oikos* **35**: 230–235.

Paige, K. N., and T. G. Whitham (1985). Individual and population shifts in flower color by scarlet gilia: A mechanism for pollinator tracking. *Science* **227**: 315–317.

Paine, R. T. (1966). Food web complexity and species diversity. *American Naturalist* **199**: 65–75.

Parra, R. (1978). Comparison of foregut and hindgut fermentation in herbivores. Pages 205–230 *in* G. G. Montgomery, Editor. The Ecology of Arboreal Folivores. Smithsonian Institution Press, Washington, D.C.

Pease, J. L., R. H. Vowles, and L. B. Keith (1979). Interaction of snowshoe hares and woody vegetation. *Journal of Wildlife Management* **43**: 43–60.

Pennycuick, C. J. (1979). Energy costs of locomotion and the concept of ''foraging radius.'' Pages 164–184 *in* A. R. E. Sinclair and M. Norton-Griffiths, Editors. Serengeti: Dynamics of an Ecosystem. University of Chicago Press, Chicago.

Pillemer, E. A., and W. D. Tingey (1976). Hooked trichomes: A physical plant barrier to a major agricultural pest. *Science* **193**: 492–484.

Prance, G. T., and S. A. Mori (1978). Observations on the fruits and seeds of neotropical Lecythidaceae. *Britonnia* **39**: 21–23.

Pratt, T. K., and E. W. Stiles (1985). The influence of fruit size and structure on composition of frugivore assemblages in New Guinea. *Biotropica* **17**: 314–321.

Prazmo, W. (1961). Genetic studies on the genus *Aquilegia* L. *Acta Societatis Botanicorum Poloniae* **39**: 425–442.

Price, P. W. (1984). Insect Ecology. Second Edition. John Wiley and Sons, New York.

Price, P. W., C. N. Slobodchikoff, and W. S. Gaud, Editors (1984). A New Ecology. John Wiley and Sons, New York.

Ramirez, W. (1974). Coevolution of *Ficus* and *Agaonidae*. *Annals of the Missouri Botanical Garden* **61**: 770–780.

Rausher, M. D. (1983). Ecology of host-selection behavior in phytophagous insects. Pages 223–258 *in* R. F. Denno and M. McClure, Editors. Variable Plants and Herbivores in Natural and Managed Systems. Academic Press, New York.

Real, L. A. (1981). Uncertainty and pollinator–plant interactions: The foraging behavior of bees and wasps at artificial flowers. *Ecology* **62**: 200–26.

Real, L. A., Editor (1983). Pollination Biology. Academic Press, New York.

Real, L. A., and T. Caraco (1986). Risk and foraging in stochastic environments: Theory and evidence. *Annual Review of Ecology and Systematics* **17**: 371–390.

Regal, P. J. (1977). Ecology and evolution of flowering plant dominance. *Science* **196**: 622–629.

Regal, P. J. (1982). Pollination by wind and animals: Ecology of geographic patterns. *Annual Review of Ecology and Systematics* **13**: 497–524.

Rehr, S. S., P. P. Feeny, and D. H. Janzen (1973). Chemical defense in Central American non-ant-acacias. *Journal of Animal Ecology* **42**: 405–416.

Reichardt, P. B., J. P. Bryant, T. P. Clausen, and G. D. Wieland (1984). Defense of winter-dormant Alaska paper birch against snowshoe hares. *Oecologia* **65**: 58–69.

Rhoades, D. F. (1983). Herbivore population dynamics and plant chemistry. Pages 155–220 *in* R. F. Denno and M. McClure, Editors. Variable Plants and Herbivores in Natural and Managed Systems. Academic Press, New York.

Rhoades, D. F. (1985). Offensive–defensive interactions between herbivores and plants: Their relevance in herbivore population dynamics and ecological theory. *American Naturalist* **125**: 205–238.

Rhoades, D. F., and R. G. Cates (1976). Toward a general theory of plant antiherbivore chemistry. *Recent Advances in Phytochemistry* **19**: 168–213.

Rick, C. M., and R. I. Bowman (1961). Galapagos tortoises. *Evolution* **15**: 407–417.

Robbins, C. T. (1983). Wildlife Feeding and Nutrition. Academic Press, New York.

Robbins, C. T., T. A. Hanley, A. E. Hagerman, O. Hjeljord, D. L. Baker, C. C. Schwartz, and W. W. Mautz (1987). Role of tannins in defending plants against ruminants; reduction in protein availability. *Ecology* **68**: 98–107.

Robinson, T. (1979). The evolutionary ecology of alkaloids. Pages 413–448 *in* G. A. Rosenthal and D. H. Janzen, Editors. Herbivores: Their Interaction with Secondary Plant Metabolites. Academic Press, New York.

Rokhsar, D. S., P. W. Anderson, and D. L. Stein (1986). Self-organization in prebiological systems: Simulations of a model for the origin of genetic information. *Journal of Molecular Evolution* **23**: 119–126.

Romer, A. S. (1959). The Vertebrate Story. University of Chicago Press, Chicago.

Root, R. B. (1973). Organization of a plant–arthropod association in simple and diverse habitats: The fauna of collards *(Brassica oleracea)*. *Ecological Monographs* **43**: 95–124.

Rosenthal, G. A. (1977). The biological effects and mode of action of L-cananavine, a structural analogue of L-arginine. *Quarterly Review of Biology* **52**: 155–178.

Rosenthal, G. A., and D. H. Janzen, Editors. (1979). Herbivores: Their Interaction with Secondary Plant Metabolites. Academic Press, New York.

Rosenthal, G. A., C. G. Hughes, and D. H. Janzen (1982). L-Canavanine, a dietary nitrogen source for the seed predator *Caryedes brasiliensis* (Bruchidae). *Science* **217**: 353–355.

Roughgarden, J. (1983). The theory of coevolution. Pages 33–64 *in* D. Futuyma and M. Slatkin, Editors. Coevolution. Sinauer Associates, Sunderland, Massachusetts.

Ryan, C. A. (1979). Proteinase inhibitors. Pages 599–618 *in* G. A. Rosenthal and D. H. Janzen, Editors. Herbivores: Their Interaction with Secondary Plant Metabolites. Academic Press, New York.

Ryan, C. A. (1983). Insect-induced chemical signals regulating natural plant protection responses. Pages 43–60 *in* R F. Denno and M. McClure, Editors. Variable Plants and Herbivores in Natural and Managed Systems. Academic Press, New York.

Schemske, D. W. (1980). The evolutionary significance of extrafloral nectar production by *Costus woodsonii* (Zingiberaceae): An experimental analysis of ant protection. *Journal of Animal Ecology* **68**: 959–967.

Schemske, D. W. (1983). Limits to specialization and coevolution in plant–animal mutualisms. Pages 67–109 *in* M. H. Nitecki, Editor. Coevolution. University of Chicago Press, Chicago.

Schemske, D. W., and C. C. Horvitz (1984). Variation among floral visitors in pollination ability: A precondition for mutualism specialization. *Science* **225**: 519–521.

Schemske, D. W., M. F. Willson, M. N. Melampy, L. J. Miller, L. Verner, K. M. Schemske, and L. B. Best (1978). Flowering ecology of some spring woodland herbs. *Ecology* **59**: 351–366.

Schupp, E. W. (1986). *Azteca* protection of *Cecropia:* Ant occupation benefits juvenile trees. *Oecologia* **70**: 379–385.

Scriber, J. M. (1983). Evolution of feeding specialization, physiological efficiency, and host races in selected Papilionidae and Saturnidae. Pages 373–412 *in* R. F. Denno and M. McClure, Editors. Variable Plants and Herbivores in Natural and Managed Systems. Academic Press, New York.

Scriber, J. M. (1984). Host–plant suitability. Pages 159–204 *in* W. J. Bell and R. T. Carde, Editors. Chemical Ecology of Insects. Chapman and Hall, New York.

Sherman, I. W., and V. G. Sherman (1983). Biology: A Human Approach. Oxford University Press, New York.

Siegel, S. (1956). Non-parametric Statistics. McGraw-Hill, New York.

Sinclair, A. R. E., and M. Norton-Griffiths (1979). Serengeti: Dynamics of an Ecosystem. University of Chicago Press, Chicago.

Singer, M. C. (1982). Quantification of host preference by manipulation of oviposition behavior in the butterfly *Euphydryas editha. Oecologia* **52**: 224–229.

Slama, K. (1979). Insect hormones and antihormones in plants. Pages 683–706 *in* G. A. Rosenthal and D. H. Janzen, Editors. Herbivores: Their Interaction with Secondary Plant Metabolites. Academic Press, New York.

Slansky, F., Jr., and P. Feeny (1977). Stabilization of the rate of nitrogen accumulation by larvae of the cabbage butterfly on wild and cultivated food plants. *Ecology* **47**: 209–228.

Smallwood, J. (1982). The effect of shade and competition on emigration rate in the ant *Aphaenogaster rudis. Ecology* **63**: 124–134.

Smallwood, J., and D. C. Culver (1979). Colony movements of some North American ants. *Journal of Animal Ecology* **48**: 373–382.

Smith, C. C., and D. Follmer (1972). Food preferences of squirrels. *Ecology* **53**: 82–91.

Smythe, N., W. Glanz, and E. G. Leigh, Jr. (1982). Population regulation in some terrestrial frugivores. Pages 277–238 *in* E. G. Leigh, Jr., A. S. Rand, and D. M. Windsor, Editors. The Ecology of a Tropical Forest: Seasonal Rhythms and Long-term Changes. Smithsonian Press, Washington, D.C.

Snow, D. W. (1961). The natural history of the oilbird, *Steatornis caripensis,* in Trinidad, West Indies. Part 1. *Zoologica* **46**: 27–47.

Snow, D. W. (1971). Evolutionary aspects of fruit-eating in birds. *Ibis* **113**: 194–202.

Snow, D. W. (1981). Tropical frugivorous birds and their food plants: A world survey. *Biotropica* **13**: 1–14.

Sorensen, A. E. (1986). Seed dispersal by adhesion. *Annual Review of Ecology and Systematics* **17**: 443–464.

Sork, V. L. (1985). Germination response in a large-seeded neotropical tree species, *Gustavia superba* (Lecythidaceae). *Biotropica* **17**: 130–136.

Sowls, L. K. (1984). The Peccaries. University of Arizona Press, Tucson.

Sporne, K. R. (1965). The Morphology of Gymnosperms: The Structure and Evolution of Primitive Seed-plants. Hutchinson University Library, London.

Stanton, M. L., A. A. Snow, and S. N. Handel (1986). Floral evolution: Attractiveness to pollinators increases male fitness. *Science* **232**: 1625–1627.

Stebbins, G. L. (1974). Flowering Plants: Evolution Above the Species Level. Harvard University Press, Cambridge, Massachusetts.

Stebbins, G. L. (1982). Perspectives in evolutionary time. *Evolution* **36**: 1109–1118.

Steenbergh, W. F., and C. W. Lowe (1977). Ecology of the saguaro: II. National Park Service Scientific Monograph No. 8, Washington, D.C.

Stephenson, A. G. (1981). Flower and fruit abortion: Proximate causes and ultimate functions. *Annual Review of Ecology and Systematics* **12**: 253–279.

Stewart, W. N. (1983). Paleobotany and the Evolution of Plants. Cambridge University Press, Cambridge.

Stiles, F. G. (1975). Ecology, flowering phenology, and hummingbird pollination of some Costa Rican *Heliconia* species. *Ecology* **56**: 285–301.

Storer, T. I. (1951). General Zoology. McGraw-Hill, New York.

Strickler, K. (1979). Specialization and foraging efficiency of solitary bees. *Ecology* **60**: 998–1009.

Strong, D., J. H. Lawton, and R. Southwood (1984). Insects on plants. Blackwell Scientific, Oxford.

Swain, T. (1978). Plant–animal coevolution; a synoptic view of the Paleozoic and mesozoic. Pages 3–20 *in* J. B. Harborne, Editor. Biochemical Aspects of Plant and Animal Coevolution. Academic Press, New York.

Swenson, M. J., Editor. (1977). Duke's Physiology of Domestic Animals. Tenth Edition. Cornell University Press, Ithaca.

Temple, S. A. (1977). Plant–animal mutualism: Coevolution with dodo leads to near extinction of plant. *Science* **197**: 885–886.

Temple, S. A. (1979). The dodo and the tambalacoque tree. *Science* **203**: 1364.

Templeton, A. R., and L. E. Gilbert (1985). Population genetics and the coevolution of mutualisms. Pages 128–144 *in* D. Boucher, Editor. The Biology of Mutualisms: Ecology and Evolution. Oxford University Press, Oxford.

Terborgh, J. (1983). Five New World Primates: A Study in Comparative Ecology. Princeton University Press, Princeton.

Thompson, J. N. (1982). Interaction and Coevolution. John Wiley and Sons, New York.

Thompson, J. N. (1986). Patterns in coevolution. Pages 119–143 *in* A. R. Stone and D. L. Hawksworth, Editors. Coevolution and Systematics. Clarendon Press, Oxford.

Thomson, J. D., W. P. Maddison, and R. C. Plowright (1982). Behavior of bumble bee pollinators of *Aralia hispida* Vent. (Araliaceae). *Oecologia* **54**: 326–336.

Thomson, J. D., M. V. Price, N. M. Waser, and D. A. Stratton (1986). Comparative studies of pollen and fluorescent dye transport by bumble bees visiting *Erythronium grandiflorum. Oecologia* **69**: 561–566.

Tiffney, B. H. (1977). Fossil angiosperm fruits and seeds. *Journal of Seed Technology* **2**(2): 54–71.

Tiffney, B. H. (1981). Diversity and major events in the evolution of land plants. Pages 193–230 *in* K. J. Niklas, Editor. Paleobotany, Paleoecology, and Evolution. Volume II. Praeger Publishers, New York.

Tilman, D. (1978). Cherries, ants and tent caterpillars: Timing of nectar production in relation to susceptibility of caterpillars to ant predation. *Ecology* **59**: 686–692.

Troyer, K. (1982). Transfer of fermentative microbes between generations in a herbiv-orous lizard. *Science* **216:** 540–542.

Troyer, K. (1984a). Structure and function of the digestive tract of a herbivorous lizard *Iguana iguana. Physiological Zoology* **57:** 1–8.

Troyer, K. (1984b). Diet selection and digestion in *Iguana iguana:* The importance of age and nutrient requirements. *Oecologia* **61:** 201–207.

Turner, J. R. G. (1981). Adaptation and evolution in *Heliconius:* A defense of neodar-winism. *Annual Review of Ecology and Systematics* **12:** 99–121.

Turner, R. M., S. M. Alcorn, and G. Olin (1969). Mortality of transplanted saguaro seedlings. *Ecology* **50:** 835–844.

Vander Wall, S. B., and R. P. Balda (1977). Coadaptations of the Clark's nutcracker and the pinyon pine for efficient seed harvest and dispersal. *Ecological Mono-graphs* **47:** 89–111.

van der Pijl, L. (1972). Principles of Dispersal in Higher Plants. Second Edition. Sprin-ger-Verlag, Berlin.

Van Soest, P. J. (1982). Nutritional Ecology of the Ruminant. O & B Books, Corvallis, Oregon.

Vesey-FitzGerald, D. F. (1960). Grazing succession among East African game. *Journal of Mammalogy* **41:** 161–171.

Via, S. (1984). The quantitative genetics of polyphagy in an insect herbivore. I. Genotype-environment interaction in larval performance on different host plant species. *Evolution* **38:** 881–895.

Via, S., and R. Lande (1985). Genotype-environment interaction and the evolution of phenotypic plasticity. *Evolution* **39:** 505–522.

Vitousek, P. M., and R. L. Sanford, Jr. (1986). Nutrient cycling in moist tropical forest. *Annual Review of Ecology and Systematics* **17:** 137–168.

Vogels, G. D., W. F. Hoppe, and C. K. Stumm (1980). *Applied and Environmental Microbiology* **40:** 608–612.

von Frisch, K. (1953) The Dancing Bees. Harcourt, Brace and Company, New York.

Waddington, K. D. (1981). Factors influencing pollen flow in bumblebee-pollinated *Delphinium virescens. Oikos* **37:** 153–159.

Waddington, K. D. (1983). Foraging behavior of pollinators. Pages 213–239 *in* L. Real, Editor. Pollination Biology. Academic Press, New York.

Walsberg, G. E. (1975). Digestive adaptations of *Phainopepla nitens* associated with the eating of mistletoe berries. *Condor* **77:** 169–174.

Waser, N. M. (1983). The adaptive nature of floral traits: Ideas and evidence. Pages 242–286 *in* L. Real, Editor. Pollination Biology. Academic Press, New York.

Webb, S. D. (1983). The rise and fall of the late Miocene ungulate fauna in North America. Pages 267–306 *in* M. H. Nitecki, Editor. Coevolution. University of Chicago Press, Chicago.

Webb, S. D. (1984). Ten million years of mammal extinctions in North America. Pages 189–210 *in* P. S. Martin and R. G. Klein, Editors. Quaternary Extinctions: A Prehistoric Revolution. University of Arizona Press, Tucson.

Wheeler, W. M. (1942). Studies on neotropical ant-plants and their ants. *Bulletin of the Museum of Comparative Zoology* **90:** 1–263.

Wheelwright, N. T. (1985). Fruit size, gape width, and the diets of fruit-eating birds. *Ecology* **66:** 808–818.

Wheelwright, N. T., and G. Orians (1982). Seed dispersal by animals: Contrasts with pollen dispersal, problems of terminology, and constraints on coevolution. *American Naturalist* **119:** 402–413.

White, S. C. (1974). Ecological Aspects of Growth and Nutrition in Tropical Fruit-eating Birds. Ph.D. Dissertation, University of Pennsylvania, Philadelphia.

Wickler, W. (1968). Mimicry in Plants and Animals. McGraw-Hill, New York.

Wiebes, J. T. (1979). Co-evolution of figs and their pollinators. *Annual Review of Ecology and Systematics* **10**: 1–12.

Wieczorek, H. (1976). The glycoside receptor of the larvae of *Mamestra brassicae* L. (Lepidoptera, Noctuidae). *Journal of Comparative Physiology* **106**: 153–176.

Williams, G. (1975). Sex and Evolution. Princeton University Press, Princeton.

Williams, W. G., G. G. Kennedy, R. T. Yamamoto, J. D. Thacker, and J. Bordner (1980). 2-Tridecanone: A naturally occurring insecticide from the wild tomato *Lycopersicon hirsutum* f. *glabratum. Science* **207**: 888–889.

Willson, M. F. (1983). Plant Reproductive Biology. John Wiley and Sons, New York.

Wolf, L. L., F. R. Hainsworth, and F. G. Stiles (1973). Energetics of foraging: Rate and efficiency of nectar extraction by hummingbirds. *Science* **176**: 1351–1352.

Wolin, C. L. (1985). The population dynamics of mutualistic systems. Pages 248–269 *in* D. H. Boucher, Editor. The Biology of Mutualism: Ecology and Evolution. Oxford University Press, New York.

Worthington, A. (1982). Population sizes and breeding rhythms of two species of manakins in relation to food supply. Pages 343–430 *in* E. G. Leigh, Jr., A. S. Rand, and D. M. Windsor, Editors. The Ecology of a Tropical Forest: Seasonal Rhythms and Long-term Changes. Smithsonian Institution Press, Washington, D.C.

Wright, S. (1946). Isolation by distance under diverse systems of mating. *Genetics* **31**: 39–59.

Wright, S. (1980). Genic and organismic selection. *Evolution* **34**: 825–843.

Young, T. P. (1984). The comparative demography of semelparous *Lobelia telekii* and iteroparous *Lobelia keniensis* on Mount Kenya. *Journal of Ecology* **72**: 637–650.

Young, T. P. (1985). *Lobelia telekii* herbivory, mortality, and size at reproduction: Variation with growth rate. *Ecology* **66**: 1879–1883.

Young, T. P., and A. P. Smith. Alpine herbivory on Mount Kenya. *In* P. Rundel, Editor. Tropical Alpine Environments: Plant Form and Function. Springer-Verlag, Berlin. In press.

Zar, J. (1984). Biostatistical Analysis. Second Edition. Prentice-Hall, Englewood Cliffs, New Jersey.

Index